D0606226

BIRDS *of* SOUTHERN CALIFORNIA'S DEEP CANYON

Ladder-backed Woodpecker (*Picoides scalaris*)

BIRDS *of* SOUTHERN CALIFORNIA'S DEEP CANYON

Color photographs by Debra and Wes Weathers

Wesley W. Weathers

UNIVERSITY OF CALIFORNIA PRESS
Berkeley, Los Angeles, London

University of California Press
Berkeley and Los Angeles, California

University of California Press, Ltd.
London, England

Library of Congress Cataloging in Publication Data

Weathers, Wesley W.
Birds of Southern California's Deep Canyon.

Includes bibliographical references and index.
1. Birds—California—Deep Canyon. 2. Birds—California—Coachella Valley. I. Weathers, Debra. II. Title.
QL684.C2W43 1982 598.29794'97 82-13382
ISBN 0-520-04754-0

Printed in the United States of America

1 2 3 4 5 6 7 8 9

CONTENTS

FOREWORD

This is the fifth in a series of books that are reports on research conducted at or near the Philip L. Boyd Deep Canyon Desert Research Center. The preceding books in the series include: *Mammals of Deep Canyon*, by R. Mark Ryan (1968); *Ants of Deep Canyon*, by G. C. and Jeanette Wheeler (1973); *Deep Canyon, A Desert Wilderness for Science*, edited by Irwin P. Ting and Bill Jennings (1976); and *Plants of Deep Canyon*, by Jan G. Zabriskie (1979).

The Boyd Research Center is a 6,727 hectare reserve that is located near Palm Desert, Riverside County, California. It is one of twenty-six reserves maintained by the University of California in its Natural Land and Water Reserves System. The Boyd Research Center is administered by the Riverside campus of the University of California.

The geographic area utilized for the investigation of the birds of Deep Canyon, as in the earlier studies mentioned above, was considerably larger than the Boyd Research Center itself. The entire study site is called the Deep Canyon Transect. It includes the Deep Canyon region of the Santa Rosa Mountains and a portion of the adjacent Coachella Valley in the Colorado Desert (a subdivision of the Sonoran Desert). The study site is a rectangle 34 kilometers long by 19 kilometers wide that includes the Boyd Research Center. The Deep Canyon Transect ranges in elevation from 9 meters on the floor of the Coachella Valley to 2,657 meters at the top of Toro Peak.

Few other places in the world offer the unusual opportunity for one to be basking in winter at air temperatures in the high seventies (Fahrenheit), while several feet of snow are lying on the ground less than 32 airline kilometers away. This elevational gradient allows nearly every habitat that occurs in inland southern California to be represented within the boundaries of the Transect. Therefore, it is not overly surprising that 217 species of birds, out of the approximately 500 species that are known to occur in southern California, have been recorded on the Deep Canyon Transect to date.

This book is intended for two audiences: (1) The interested layman can learn which species of birds occur in the Coachella Valley and adjacent

Santa Rosa Mountains. The information available includes the types of habitats in which each species can be found, as well as the time of year the birds are present. Although the book is not intended to be a field guide, the colored plates certainly will help identify some of the more common birds of this portion of southern California. (2) The desert-oriented scientist will also find considerable unpublished data regarding the population dynamics of this region's bird communities. Much of the book deals with the energy flow through the birds in different habitats. These animals represent a major portion of the complex distribution of energy within these habitats. Therefore, this information will be useful to anyone interested in the energetics of wild populations.

Wilbur W. Mayhew, Director
Philip L. Boyd Deep Canyon
Desert Research Center
Palm Desert, California

ACKNOWLEDGMENTS

Science is a social enterprise, rather than a solitary endeavor, and I owe considerable gratitude to the many people who contributed to this book. Chief among these is Wilbur W. Mayhew. Bill encouraged me to undertake this project, generously shared his treasury of original data on bird densities, and critically read the entire manuscript, including the penultimate version. He is largely responsible for the book's depth. Special thanks are due to my wife, Debra. She made life in the field pleasant, helped census birds, spent many uncomfortable hours crouched in the photography blind, analyzed most of the raw data, and helped untwist my prose. Philip and Dorothy Boyd were early and enthusiastic supporters, providing help ranging from personal to financial encouragement. Jan Zabriskie's friendship, hospitality, and encouragement helped me see the project through to its completion.

The staff of the Living Desert Reserve, especially Karen A. Sausman (Director), Art Carrillo, Sue Fuller, and Fred LaRue, provided help ranging from assistance with photography to warm friendship. Their knowledge of the area's birds helped resolve many questions regarding species' ranges.

I am particularly grateful to those who commented on the manuscript, or parts of it, in various stages of writing: Eugene Cardiff, Page Frechette, Bill Jennings, and Ned K. Johnson. The editorial comments of Bill Mayhew, Jan Zabriskie, and my wife, Debra, helped clarify much of the prose. Richard E. MacMillen devoted considerable energy to reading critically the final manuscript. Andrew Engilis prepared the range maps and Carol Shapiro drew many of the graphs. Karen Sausman and George Lepp's many helpful suggestions constituted a crash course in bird photography, and Frances Fraser White's generous gift of photographic equipment helped make the photographs possible.

The research reported here was carried out under the auspices of the Philip L. Boyd Deep Canyon Desert Research Center. Additional support was provided by the National Science Foundation (grants PCM

76-18314 and DEB 80-22765) and the College of Agriculture, University of California, Davis. This is contribution number 6 of the University of California Natural Land and Water Reserves System.

one

THE BACKGROUND

My original interest in desert birds was concerned with their physiological adaptations to heat and aridity. By simulating desert conditions in the laboratory and then carefully measuring body temperature, oxygen consumption, and evaporative water loss, I hoped to discover how some desert birds survive without drinking water. The University of California's field station at Deep Canyon provided a convenient base for my collecting trips into adjacent areas, and as I spent time there my interest in birds gradually broadened. I became fascinated by the character and variety of Deep Canyon's bird communities, which, because of a steep climatic gradient, are compressed into a short linear distance—hot, dry desert and cool coniferous forest a mere 18 km apart. I was frustrated, however, to find that little information existed about these communities. Definite answers could not even be found to such basic questions as what birds are found at Deep Canyon, how many of them are there, and where and when do they occur. This surprised me since the field station (the Philip L. Boyd Deep Canyon Desert Research Center) is a focal point for ecological studies in southern California. Clearly, here was an opportunity to expand our knowledge of birds by simple observation.

THE SETTING

Deep Canyon is an ecologically important link between the Colorado Desert—a subdivision of the international Sonoran Desert—and the Santa Rosa Mountains—a part of the international Peninsular Range. Located at the northwestern corner of southern California's Coachella Valley (fig. 1), Deep Canyon cleaves the north slope of the Santa Rosa Mountains. Rugged and starkly beautiful, the canyon's 394 m high cliffs make it one of the more spectacular gorges draining the slopes of the contiguous Santa Rosa and San Jacinto mountains (fig. 2). The canyon's intermittent stream, Deep Canyon Creek, carries runoff from the top of

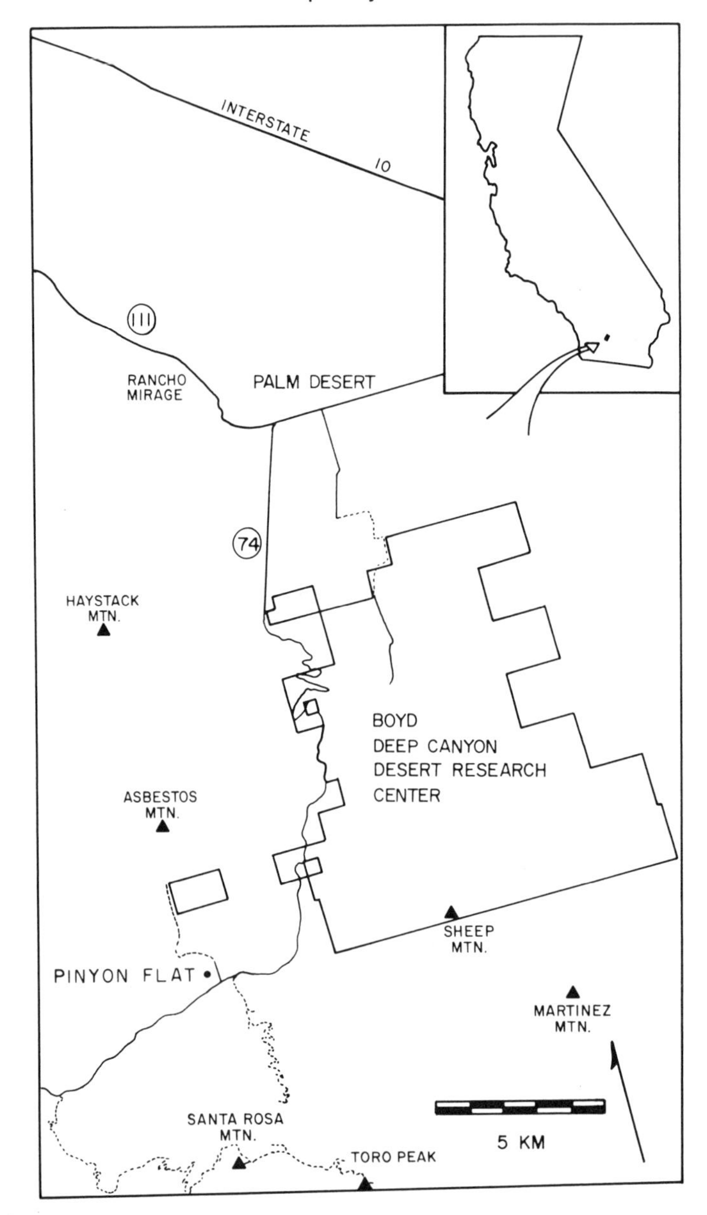

FIGURE 1 Map showing location of the Deep Canyon Transect.

the Santa Rosa Mountains to the floor of the Coachella Valley 2600 m below.

Because climate changes abruptly with elevation, a journey from the Coachella Valley to the top of the Santa Rosa Mountains is ecologically equivalent to a latitudinal journey from San Diego to Edmonton, Alberta. Ascending the Santa Rosa Mountains, one passes through a series of

FIGURE 2 The view north from atop Santa Rosa Mountain. Deep Canyon cuts across the photograph from left to right and opens onto the alluvial plain and the Coachella Valley. Sugarloaf Mountain (foreground) and Black Hill (background) rise from the plateau to the left of Deep Canyon. In the distance, the Little San Bernardino Mountains mark the boundary between the Colorado Desert and the Mojave Desert.

habitats (fig. 3) that differ in topography, climate, plant cover, and the kinds of plants and animals present. Consequently, many types of birds can be seen within a single day at Deep Canyon. Indeed, a California birding party saw an all-time record of 227 bird species during a single day by traveling along a similar gradient from the Colorado Desert across the Peninsular Range to the Pacific Ocean (Small 1974).

During the ascent of the Santa Rosa Mountains, one encounters striking differences in vegetation. Percent plant cover, for example, increases over twelve-fold from the base of the rocky slopes, reaching a maximum at the boundary between the chaparral and the coniferous forest (fig. 4). A change in vegetation type is also notable. Drought-deciduous and succulent plants of the lower elevations give way to winter-deciduous and evergreen plants higher on the mountain (fig. 5). Such a pronounced change in habitats over a short, linear distance results in a dramatic change in the bird communities.

Most of Deep Canyon's habitats appear as conspicuous altitudinal belts on the mountains' face. But one—the streamside—lacks the altitudinal belting of the others. Born of streamflow and distinguished by its mesic

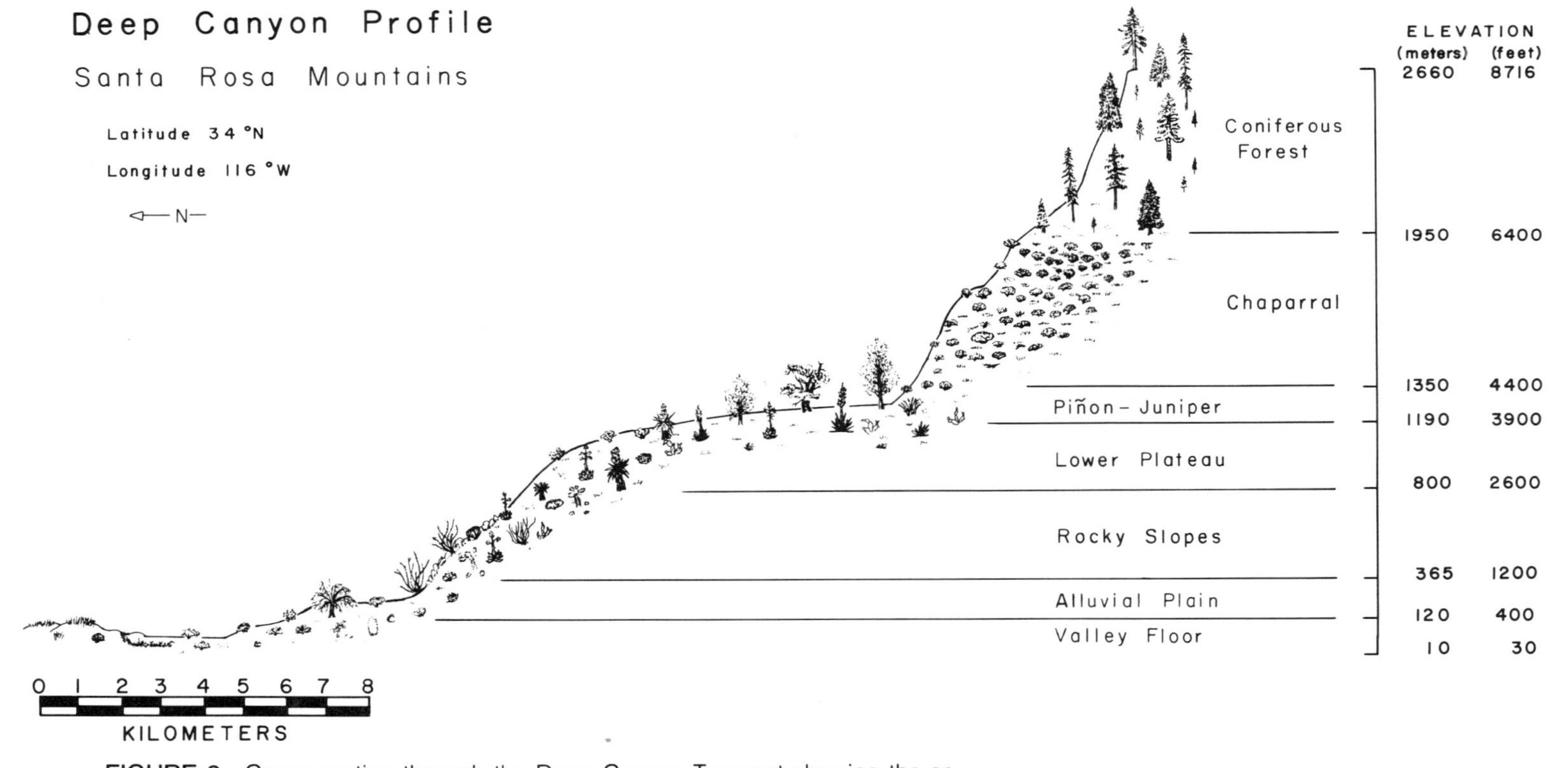

FIGURE 3 Cross section through the Deep Canyon Transect showing the sequence of habitats from the Coachella Valley floor to the top of Toro Peak.

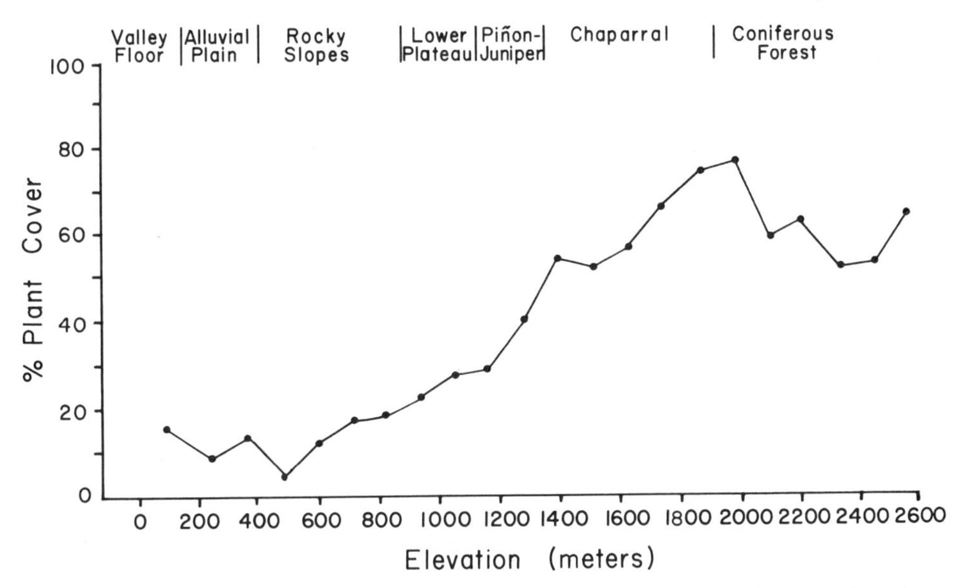

FIGURE 4 Relation of plant cover to elevation at Deep Canyon (data from Zabriskie 1979).

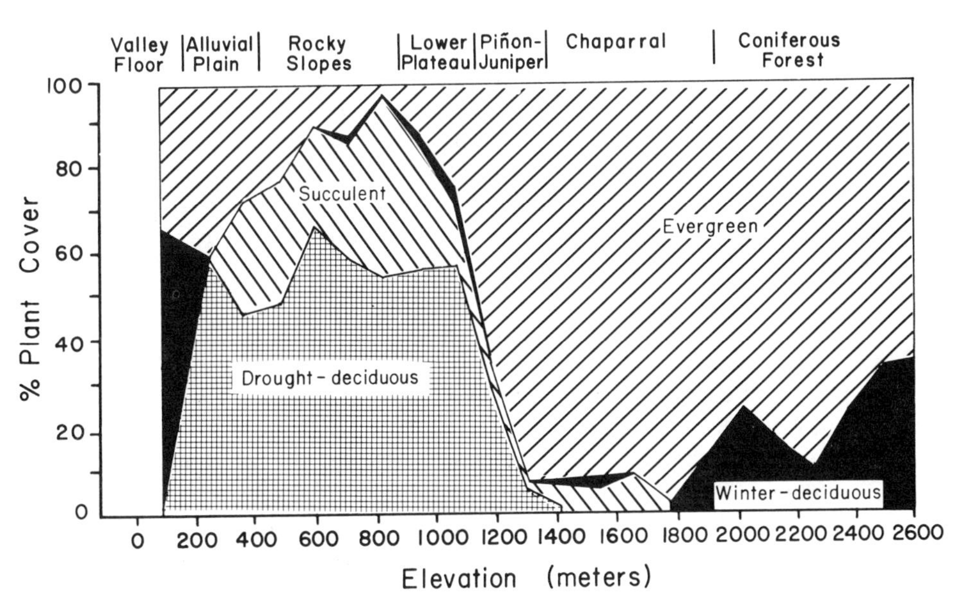

FIGURE 5 Change in growth-form of perennial plants with elevation at Deep Canyon. Note the abrupt change on the lower plateau (data from Zabriskie 1979).

condition and comparatively lush vegetation, it passes through virtually all the other habitats. Where stream courses emerge from the mountains, they widen into sandy washes, a discrete subhabitat of the alluvial plain.

Life Zones

Changes in communities with elevation are observed in mountains throughout the world and were first described in terms of life zones by C. Hart Merriam (1898). The life zone system suffers from several shortcomings, and although its use was *de rigueur* in early ecological studies, it has fallen from favor and is scarcely mentioned in many recent texts (e.g., Ricklefs 1973, Pielou 1974, Krebs 1978). For the mountains of western North America, however, it remains a serviceable basis for faunal analyses (see Lowe 1964 for additional insights). Although this book mainly employs topographical habitats to describe communities, life zones must be mentioned because of their continued use by western biologists and because of their importance in earlier studies at Deep Canyon (e.g., Grinnell and Swarth 1913, Ryan 1968).

The Deep Canyon Transect encompasses three classic life zones—Lower Sonoran, Upper Sonoran, and Transition. The Lower Sonoran encompasses the desert habitats from the valley floor through the lower plateau and continues upward through an ecotone of mixed yucca, juniper, agave, and cactus. It embraces several associations of xerophylic (literally "dry-loving") plant species characteristic of the creosote bush (*Larrea tridentata*)* scrub community. The Upper Sonoran corresponds to the piñon-juniper and chaparral habitats, which occur between 1065 and 1956 m elevation. The highest zone, the Transition, is represented by a forest of jeffrey pine (*Pinus jeffreyi*).

Many of Deep Canyon's 112 nesting species show considerable life zone fidelity, with the Transition supporting a bird community much different from that found in the adjacent Upper Sonoran, which, in turn, differs from the Lower Sonoran. This can be illustrated by listing the preferred habitat for the more common resident species (table 1).

METHODS

I spent 198 days at Deep Canyon observing, photographing, and censusing birds between March 1977 and December 1980. I walked strip transects (established by W. W. Mayhew), used tape recorded bird calls to find uncommon species, and spent hundreds of hours just observing birds. I also assembled observations made at Deep Canyon by competent bird watchers since the Research Center's establishment in 1959 (see Appendix I). From this information a clear image emerged of the occurrence and distribution of most of Deep Canyon's 217 species. Some species remain enigmatic and these are indicated by dashed lines in Appendix I.

Several published accounts of southern California's birds provided me with hints of what birds to expect at Deep Canyon. Grinnell and

*Except for those birds listed in Appendix I, scientific names are given following a species' first occurrence in the text.

TABLE 1 Habitat in Which the Species Reaches Its Greatest Density

Lower Sonoran	Upper Sonoran
Valley floor	*Piñon-juniper*
Burrowing Owl	Mourning Dove
Red-tailed Hawk	Common Raven
Say's Phoebe	Scrub Jay
Starling	Common Flicker
Alluvial plain: scrubland	Chipping Sparrow
None	Rufous-sided Towhee
Alluvial plain: desert wash	Plain Titmouse
House Finch	Piñon Jay
Gambel's Quail	*Chaparral*
Costa's Hummingbird	Lesser Goldfinch
Verdin	Bushtit
Black-tailed Gnatcatcher	Bewick's Wren
Cactus Wren	California Thrasher
Mockingbird	Mountain Quail
Rocky slopes	Wrentit
Rock Wren	Anna's Hummingbird
Loggerhead Shrike	Cooper's Hawk
Roadrunner	Brown Towhee
Lower plateau	Sage Sparrow
Black-throated Sparrow	
White-throated Swift	**Transition**
American Kestrel	*Coniferous forest*
Ladder-backed Woodpecker	Pygmy Nuthatch
Golden Eagle	Mountain Chickadee
	Dark-eyed Junco
	Band-tailed Pigeon
	Steller's Jay
	Hairy Woodpecker
	White-headed Woodpecker
	White-breasted Nuthatch
	Brown Creeper

Swarth (1913) made the first systematic study of Deep Canyon's birds. They collected birds from 1 May through 5 September 1908 in the San Jacinto and Santa Rosa mountains, and consequently, their work provides a valuable historical perspective. Miller and Stebbins (1964) conducted an intensive bird survey in the Mojave Desert's Joshua Tree National Monument, located 50 km north of Deep Canyon across the Coachella Valley. Portions of the Mojave Desert resemble the middle elevations of the Deep Canyon Transect, and both places share many of the same species. Miller and Stebbins's data gave me insights into the expected arrival dates of nonresident species. Grinnell and Miller's (1944) monumental work, together with Small's (1974) recent update, are the best sources of information on the distribution of California's birds. (Garrett and Dunn's [1981] work on southern California's birds appeared after this manuscript was completed. It is an excellent supplement to Grinnell and Miller [1941].) These sources provided me with a basis for making predictions about what birds should occur at Deep Canyon. Most of the predictions were verified; some were not.

In addition to the above works on birds, information is available on the mammals (Ryan 1968), ants (Wheeler and Wheeler 1973), and plants (Zabriskie 1979) of Deep Canyon, making this area ideal for further research on community relations.

Censuses

Bird density was determined from censuses of strip transects 50 m wide. The number of transects, total area sampled, and the number of censuses per habitat are given in table 2. Most of the 684 censuses were conducted during 1979 and 1980 by Wilbur W. Mayhew and me. The rocky slopes habitat, however, was censused primarily by Barbara Carlson during 1978 through 1980 (see Carlson 1979*a, b*), with additional censuses by Mayhew in 1979 and 1980. Working separately, we used all available cues to detect birds as we advanced with frequent pauses at speeds averaging about 1.5 km/h. Like Emlen (1979), we assumed that, in these open habitats, all species occurring inside the 25 m boundary were detected. Hence, no adjustments were made in our tallies for secretive species. Census times were distributed randomly throughout the day from sunrise to sunset. This is a departure from the usual practice of limiting censuses to near sunrise. We found that in these open habitats, time of day has no significant influence on density estimates derived from strip transect censuses (Weathers and Mayhew 1981). Each habitat was censused during all months of the year, and the data from different years were combined. The most thoroughly studied habitat (desert wash) was censused, on the average, once every two-and-a-half days, while the least studied habitat (valley floor) was censused once every eight-and-a-half days.

Bird Density

From the census results, the density of each species was calculated as the number of individuals per 40 hectares; a standard unit in avian ecology. An area of 40 ha (the abbreviation for hectares) is equal to 90 football fields (end zones excluded). Thus, a density of 90 birds/40 ha is equivalent to one sparrow standing in the middle of a football field. We found many densities in this range.

In the habitat chapters, the density of each species is given together with its percent occurrence (i.e., percent of the total censuses during which the species was encountered). This latter value gives additional

TABLE 2 Distribution by Habitat of Strip Transects and Number of Censuses

Habitat	No. transects	Area censused (ha)	Total no. censuses
Valley floor	1	9.0	43
Desert wash	6	32.9	142
Scrubland	4	24.2	105
Rocky slopes	2	19.5	58
Lower plateau	2	5.4	58
Piñon-juniper	5	20.8	100
Chaparral	2	10.0	66
Coniferous forest	1	5.0	49
Streamside	3	15.0	63
TOTAL	26	141.8	684

TABLE 3 Density and Percent Occurrence of Two Warbler Species

Species	% Occurrence	Density, birds/40 ha
Wilson's Warbler	14.6	3.84
Yellow Warbler	6.3	3.86

insights into the census results. For example, although Wilson's and Yellow Warblers have similar densities, their percent occurrences differ markedly (table 3), with twice as many Yellow Warblers being found per census. This indicates that Yellow Warblers have a greater tendency to occur in flocks than do Wilson's Warblers. Indeed, Wilson's Warblers usually occur as solitary individuals, whereas Yellow Warblers typically travel in pairs or threes. Many variables affect the percent occurrence values, however, and quantitative comparisons are not practicable.

two

WEATHER AND CLIMATE

Deep Canyon and the Coachella Valley lie within rain-shadows cast by southern California's Santa Rosa, Laguna, and San Jacinto mountains. The mountains block the inland passage of winter storms from the Pacific Ocean, and their leeward regions experience a generally arid climate. This effect is amplified by the general circulation pattern of the earth's atmosphere, which produces broad belts of desert near 30° latitude on both sides of the equator. Although the rain-shadow's impact is lessened somewhat by summer rains that originate to the south, aridity is Deep Canyon's dominant climatic feature, even at the higher, cooler elevations.

Deep Canyon's generally arid climate changes markedly from the Coachella Valley to the top of the Santa Rosa Mountains. True desert conditions prevail on the valley floor with high temperatures, low rainfall, and high potential evapotranspiration making that portion of the Colorado Desert one of North America's hottest and driest regions (Shantz and Piemeisel 1924, Walter 1971). Indeed, the valley floor is climatically similar to the deserts of Africa, Asia, and the Middle East. Eighteen kilometers from the valley floor, on Toro Peak, the climate is temperate: warm summers alternate with cold, snowy winters. On balmy spring days when valley floor temperatures are mild, snow flurries may swirl about Toro Peak. Obviously, any description of climate must be made with reference to elevation.

The seasonal climate at the base of the Santa Rosa Mountains (Boyd Research Center, elev. 290 m) alternates between mild winters and hot summers. From November through March, maximum air temperatures fluctuate between 18° and 23°C, and frosts rarely occur. In the absence of winter storms, calm sunny days and springlike conditions may prevail, with midday air temperatures rising to around 20°C. Although the months of April and May are consistently springlike, in some years spring may begin in early March or last until late June.

Spring at the Boyd Research Center is an enchanting time. Hot days give way to balmy evenings. A breeze begins at dusk and becomes the night's lullaby as a sliver of a new moon, hung low on the horizon, dimly lights the evening. Venus, Mars, Jupiter, and Saturn brighten the evening sky, lighting the way for migrant birds returning to their northern breeding grounds. Throughout the night, the trills of red-spotted toads (*Bufo punctatus*) and the chirps of crickets are punctuated by occasional coyote (*Canis latrans*) yelps or the low hoots of Great Horned Owls. Dawn breaks to reveal resident birds actively engaged in courtship and nest building among the blossoms of the desert's trees and shrubs.

Summer, in contrast, is a time of oppressive heat and occasionally violent rain storms. In June maximum air temperatures average around 36°C. Conditions become progressively hotter and drier during July, challenging the survival ability of resident birds. In late September or October air temperatures rapidly decline, heralding the end of summer and the beginning of a period of calm, dry, sunny days that last until the first storms of winter. Against this backdrop of the annual weather cycle, Deep Canyon's birds play out their lives in disparate ways.

WEATHER RECORDS

Data on precipitation, temperature, humidity, and wind are available only for the lower elevations of the Deep Canyon Transect—valley floor to the piñon-juniper woodland. The record period varies from 102 years on the valley floor (Indio) to 3 years at the 1,600-ft (488-m) site. A quantitative overview of these data follows (for additional information see Zabriskie 1979).

Rainfall

Precipitation at Deep Canyon is frontal during winter and convectional in summer. In winter, middle-latitude cyclones draw maritime polar air masses southward, bringing rain to the desert mainly between November and April. In spring, the thermal low over the desert intensifies as the continent warms up, and the subtropical high pressure strengthens offshore and shifts northward (Axelrod 1973). This prevents summer cyclonic storms from moving further south than about 42° north latitude. Summer convectional storms result from the invasion of tropical air masses originating in the south. They frequently involve the gradual build-up over several days of large thunderclouds that may dissipate in the evening or unleash their accumulated moisture in a sudden downpour. Although summer precipitation helps to moderate the severe desert conditions, the high-intensity rain results in proportionately less available water (because of greater runoff) than an equivalent amount of winter rain (Barbour and Major 1977). Massive tropical storms (hurricanes) occasionally reach Deep Canyon and cause severe flooding. Recent storms of this nature occurred in September 1939, September 1976, August 1977, and July 1979.

A bimodal pattern of rainfall (fig. 6) helps to distinguish the Sonoran Desert from the more northern Mojave Desert in which summer precipitation is rare. The amount of summer rainfall, however, varies locally

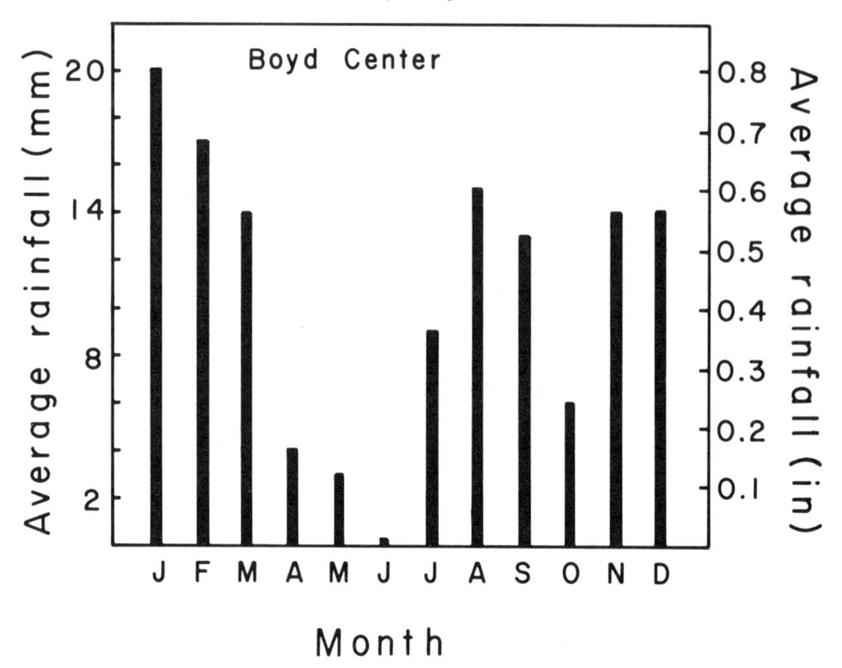

FIGURE 6 Average monthly rainfall at the Philip L. Boyd Deep Canyon Desert Research Center, 1961 through 1980.

within the Sonoran Desert. Shreve (1925) reported that it increases from about 5 percent of the annual total at the Sonoran Desert's western edge to 34 percent at the Colorado River and 50 percent at Tucson, Arizona. Conditions may have changed since 1925, as the 102-year record for Indio (near the desert's western edge) shows that one-third of the rainfall has occurred there during the summer.

Precipitation data for Deep Canyon's higher elevations are available only for the past seven years. More extensive data for Indio (102 years) and the Boyd Research Center (20 years) reveal a marked variation in annual rainfall typical of deserts. Boyd Research Center, located at the base of the Santa Rosa Mountains, is nearly 300 m higher than Indio and consequently experiences both higher rainfall and milder temperatures. From 1961 through 1980, the center averaged 129 mm of precipitation per year. Year-to-year variation in rainfall is fairly great, ranging from 0.3 to 2.6 times the annual average. Thus, over a 20-year period, the wettest years received nearly nine times as much rain as the driest years (fig. 7). The period covered by the present study (1977–1980) received above-average rainfall.

How the recent above-average rainfall may have influenced the birds of Deep Canyon is uncertain, but the sudden appearance of Zone-tailed Hawks in 1978, and their subsequent annual return, is probably due

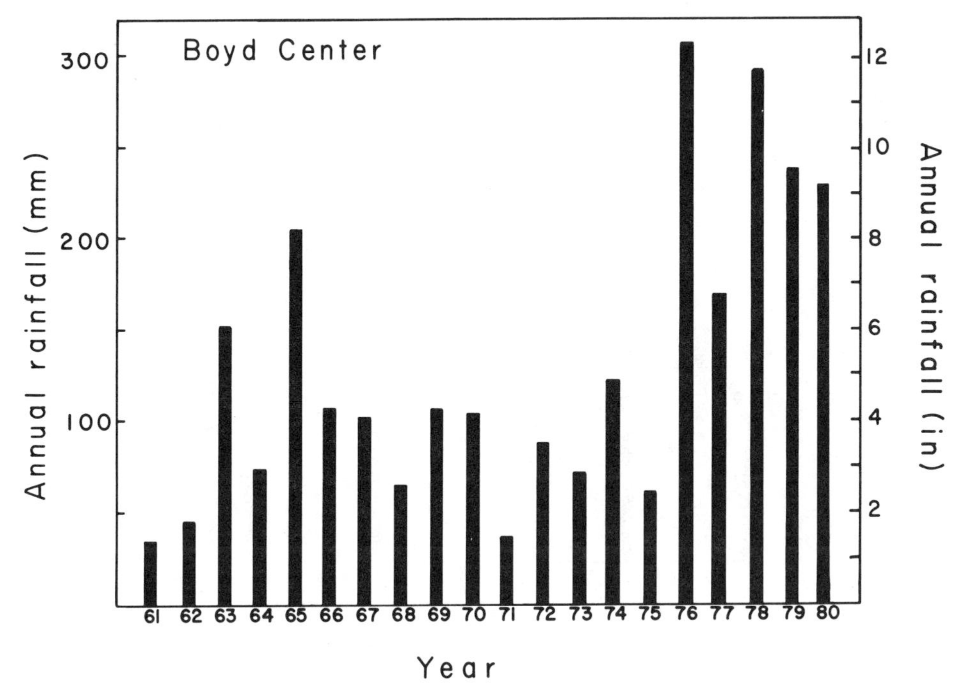

FIGURE 7 Annual rainfall at the Philip L. Boyd Deep Canyon Desert Research Center.

to wetter summers. In recent decades, several bird species have expanded their breeding ranges into southern California's montane habitats, possibly because of increased rainfall (Johnson and Garrett 1974). These include: Broad-tailed Hummingbird (*Selasphorus platycerus*), Gray Flycatcher, Virginia's Warbler (*Vermivora virginiae*), Grace's Warbler (*Dendroica graciae*), Painted Redstart (*Myioborus picta*), and Hepatic Tanager (*Piranga flava hepatica*). Of these, only the Gray Flycatcher occurs at Deep Canyon.

Because the region is mountainous, steep gradients in available moisture exist throughout Deep Canyon and the Colorado Desert. These gradients have profound effects on the region's biota and are largely responsible for the vegetational belts on the mountain slopes. Rainfall at Deep Canyon follows an elevational gradient, increasing from around 10 cm annually at Indio to 35 cm at Pinyon (Zabriskie 1979). Rainfall along the transect averaged higher than this during the years in which bird densities were determined (1979–1980) (fig. 8).

One aspect of Deep Canyon's precipitation pattern not revealed by figures 6 through 8 is its sometimes patchy nature. Summer storms in particular can result in very localized downpours. For example, during July 1977 the Taylor site received 16.8 mm of rain on a day when the

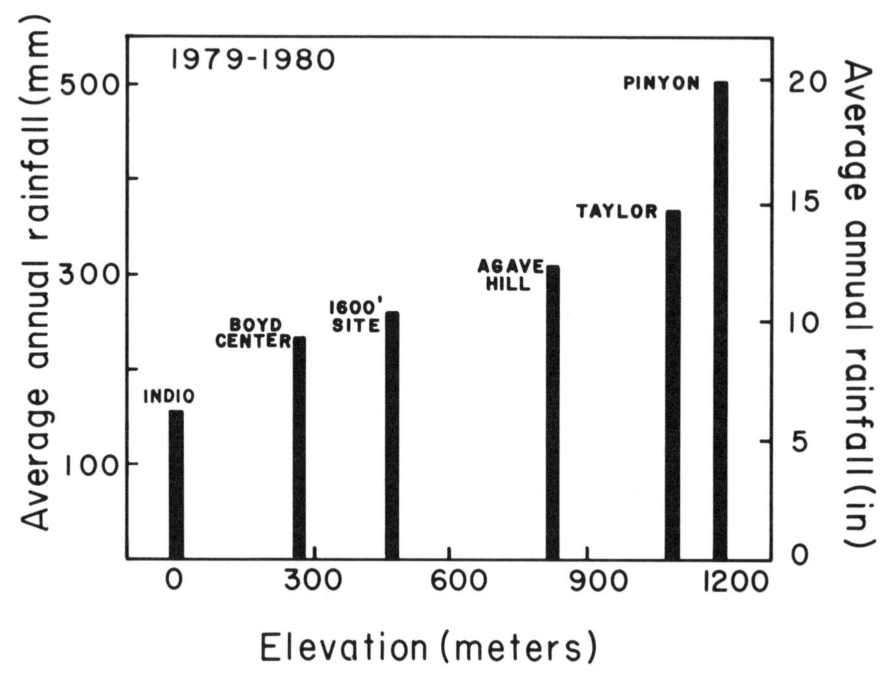

FIGURE 8 Relation of average annual rainfall to elevation at Deep Canyon during the present study.

Pinyon site (located 3.2 km away) received only 0.2 mm (Zabriskie 1979). Such localized showers may produce resource patches that birds, because of their high mobility, can exploit.

Although precipitation data provide a useful index of desert conditions, the degree of aridity is a function of both precipitation and the potential for evaporation. Precise classifications consider both (see Meigs 1953, McGinnies et al. 1968).

Temperature

Air temperatures, as measured in standard weather shelters, are available only for the lower portions of the Deep Canyon Transect. All available temperature data for the higher elevations, plus data for the past twenty years for Indio and Boyd Research Center, are summarized in figure 9.

Air temperatures on the valley floor have reached 52°C (126°F), approaching the hottest temperature recorded for the entire Colorado Desert and only slightly cooler than the absolute North American maximum (56.7°C) recorded in 1913 at Death Valley. The transect's lowest temperature (−12°C) was recorded at Pinyon. Even lower temperatures must occur during winter on Toro Peak, but no records are available.

Elevational temperature comparisons at Deep Canyon are complicated by the differing record lengths. Nevertheless, distinctive altitudinal

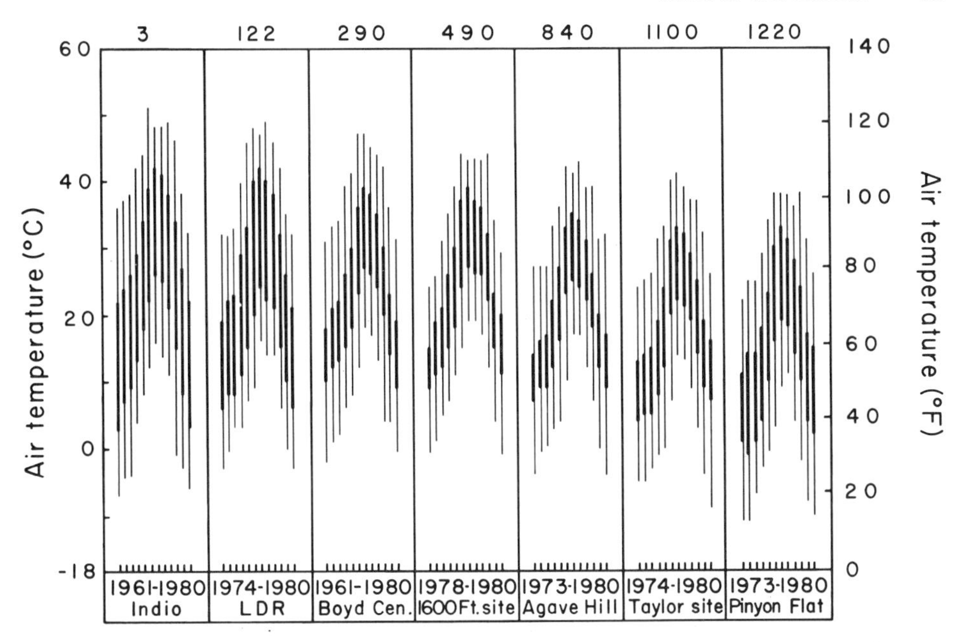

FIGURE 9 Air temperatures recorded in standard weather shelters at different sites within the Deep Canyon Transect. Site elevation (in meters) is indicated at the top of the figure. Data are plotted by month (beginning with January) and are inclusive for the indicated years. The thin vertical lines represent the absolute temperature range, and the thick vertical lines connect the mean monthly maximum and minimum temperatures. Sites are Indio, California (Indio); Living Desert Reserve, Palm Desert, California (LDR); and the Philip L. Boyd Deep Canyon Desert Research Center (Boyd Cen.).

trends in maximum and minimum air temperatures are evident (fig. 9). Maximum temperature consistently decreases with increasing elevation, while minimum temperature first increases and then decreases. Because of a nighttime temperature inversion, nights are actually several degrees warmer at the Boyd Research Center than at lower elevations. Somewhere slightly above the Boyd Research Center, minimum temperatures revert to a decreasing cline, and freezing temperatures become more common.

Air temperature, as reported here, is measured in the shade 1 to 2 m above ground. Although this is an appropriate measurement site for a large animal such as man, small birds can experience quite different conditions. A sparrow, foraging on the ground where daytime temperatures are high and wind speeds low, encounters much hotter conditions than one perched a meter above ground in a bush (fig. 10). To characterize rigorously an organism's thermal regime requires knowledge of direct and indirect solar radiation, humidity, wind velocity, and air temperature. Precise analysis of the thermal environment has proved important to understanding the abundance and distribution of species in time and space (see Gates and Schmerl 1975).

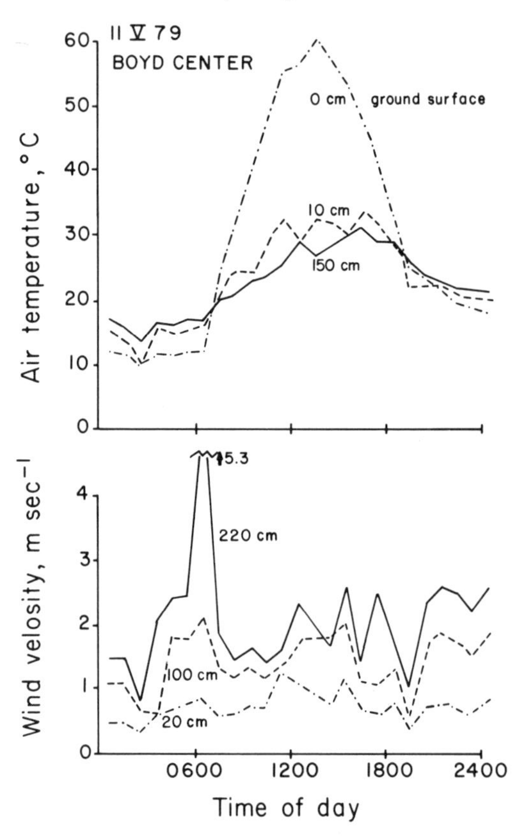

FIGURE 10 Relation of air temperature (above) and wind velocity (below), measured at different heights above the ground, to time of day (data courtesy of Dr. Park Noble, University of California, Los Angeles).

Wind

Wind is another climatic feature of the Colorado Desert that varies seasonally. Beginning in March, morning calms give way to afternoon westerly winds that increase in intensity throughout the evening, only to die out again by dawn. This pattern of almost-daily winds usually continues through summer.

Strong spring winds result from the marked temperature difference that exists between the fog-shrouded Los Angeles Basin and the sunny Coachella Valley. As valley air at ground level is heated by the sun, it becomes less dense, rises, and pulls in air and smog from the west through the San Gorgonio Pass. The pass acts like a gigantic funnel, and winds at its mouth frequently acquire sandstorm intensity. In September or October the afternoon winds gradually diminish, beginning a period of calm, sunny days that lasts until the first winter storms.

Spring's strongest winds are associated with Pacific storms that blow in from the west and south. These savage winds whip down the mountains onto the alluvial fan, sometimes last for days, and occasionally bring rain showers to the lower elevations or snow to the mountain peaks. Such storms are memorable experiences. Sudden gusts strong enough to flatten creosote bushes violently flail the branches of ocotillo (*Fouquieria splendens*) until the ground becomes a carpet of crimson blossoms. The yellow and black male Scott's Orioles are forced to forsake the wildly waving blossoms and forage lower among the flowers. Most birds are silenced by the constant roar, but if a calm briefly takes hold, White-throated Swifts can be heard twittering high overhead.

Spring winds make survival more difficult for most birds. Nests are destroyed, foraging becomes nearly impossible, and birds occasionally become impaled on the spines of cholla cactus—a hazard peculiar to the desert. Strong winds also prevent migrating species from traversing the mountains or the San Gorgonio Pass, and spectacular concentrations of birds build up on the mountains' leeward side. Desert washes become foraging sites for such normally montane species as the Western Tanager, Olive-sided Flycatcher, Western Wood Pewee, and Black-headed Grosbeak. Warblers become abundant, and a walk down a desert wash the morning after a windstorm can be a rewarding experience. For example, before a typical windstorm during May 1979, I found 50 warblers/40 ha in Coyote Wash. By the third day of the storm there were 183 warblers/40 ha (a 266 percent increase), and the number of bird species had increased from 15 to 22.

HISTORY OF DEEP CANYON'S CLIMATE

Rainfall throughout the southwestern United States has decreased since at least the Paleocene (Axelrod 1973) as a result of a general cooling trend (fig. 11). This decrease was exacerbated on the Colorado Desert during the Pleistocene (i.e., within the last 2 million years) by increased uplifting of the Sierra and Peninsular ranges and the simultaneous subsidence of the Salton Sink. Four major glacial episodes occurred during the Pleistocene, with California's mediterranean climate appearing after the first (Axelrod 1973). During the warm, dry, interglacial periods, deserts expanded throughout the southwest, only to retreat again as the ice returned and the climate became cooler and moister. The marked climatic changes during the Pleistocene—periods of glaciation alternating with warm, dry, interglacial periods—produced numerous changes in the location of Deep Canyon's habitats. Climatically induced changes in the plant communities of the southwestern deserts were paralleled by changes in the bird communities, and these are lucidly described by Hubbard (1974).

Radiocarbon dating of plant material from ancient wood rat (*Neotoma* spp.) middens reveals how the southwestern desert climate has changed during the past 22,000 years (Van Devender and Spaulding 1979). Between 22,000 and 11,000 years B.P. (before present), as the last

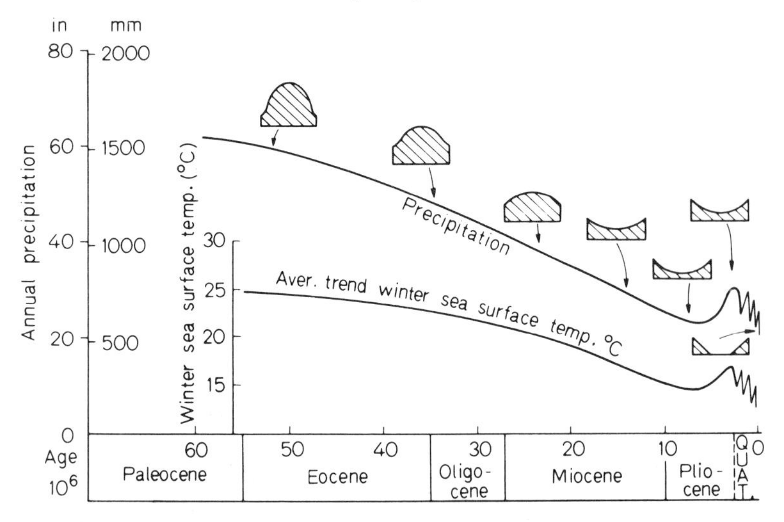

FIGURE 11 Change in California's annual precipitation and the inferred seasonal distribution of rainfall (small hatched figures) during the past 60 million years (from Axelrod 1973).

glacial period was coming to a close, the climate was mild with wet winters and cool summers producing piñon-juniper woodland in areas now occupied by desert scrub communities. By the early Holocene (11,000 to 8,000 years B.P.) the climate became drier and warmer, and the piñon-juniper woodlands were gradually replaced by middle-elevation juniper or juniper-oak woodlands, with creosote bush scrubland communities found below 300 m elevation. Then approximately 8,000 years B.P., there was a rapid, widespread, and synchronous retreat of the xeric woodlands northward and up-mountain as creosote bush-burrobush (*Ambrosia dumosa*) communities expanded into the lowlands of the Sonoran and Mojave Deserts. At about this time, palo verde (*Cercidium floridum*), saguaro (*Carnegiea giganteus*), and ironwood (*Olneya tesota*) invaded the United States from the Mexican lowlands near the head of the Gulf of California. Conditions gradually became drier, with the modern desert scrub communities becoming established only within the last 4,000 years.

The fossil record thus reveals that the present climate of the Colorado Desert is of recent origin, with juniper woodlands occupying some desert regions as recently as 8,000 years ago. Nothing is permanent. In the distant future, after our civilization has faded away, Mountain Chickadees, Piñon Jays, and Plain Titmice may once again forage in juniper woodlands where the cities of Palm Desert and Palm Springs now lie.

three

SURVEY OF BIRD COMMUNITIES

Deep Canyon's steep climatic gradient compresses several habitats into a relatively small area (see fig. 2), and each of these supports a different bird community. A detailed description of these communities and how they change seasonally is given in subsequent chapters. As a framework for the later detailed analyses, I here present an overview of the communities and consider some factors that influence their structure.

A community is a collection of populations of different species found in a particular habitat. Communities are organized along both structural and functional lines. Structural features, such as the number of species, the number of individuals per species, and when each species is found, are relatively easy to determine. Functional features, in contrast, are more complicated and difficult to quantify. For example, an uncommon species that selectively preys on nestlings will be functionally more important than its abundance (structure) suggests. Similarly, a large but uncommon insectivorous species may actually harvest more insects than a small but more common species. Knowing how species interact to divide the available resources—space, food, nest sites—is essential to understanding how communities function. But to appreciate function, we must first understand structure.

COMMUNITY STRUCTURE

Appendix I gives a broad overview of Deep Canyon's bird communities by listing the species found in each habitat. Since the species occur at different times of year, they are assigned to seasonal categories as follows: *resident species* are present throughout the year; *winter visitors* generally arrive in late fall, remain through winter and depart in spring; *summer visitors* arrive in spring, nest, and depart in fall; *migrants* merely pass through Deep Canyon during spring and/or fall. This classification scheme works well at the species level but suffers some inherent difficulties at the level of the individual. For example, although Ash-throated

Flycatchers, as a species, are summer visitors on the valley floor, many individual flycatchers seen there in spring are migrants. Because it is impossible to tell whether a given Ash-throated Flycatcher encountered during a spring census intended to remain and nest or was merely passing through, I classified them all as summer visitors. Errors resulting from faulty classifications affect mainly uncommon species and spring and fall censuses (the periods of migration). Winter and summer censuses, resident species, and common visitors are less subject to this problem. Hence, the overall community patterns are reasonably accurate.

Throughout this book, seasons refer to the following periods: winter, December through February; spring, March through May; summer, June through August; fall, September through November.

Bird Density

The number of birds found in a given area changes constantly, although less so in stable environments than in cyclic ones. In a northern hardwoods forest, bird density rose sharply from a low of about 40 birds/40 ha in winter to a June peak of around 1,200 birds/40 ha (Holmes and Sturges 1975). This cyclic pattern presumably applies to most arctic and cold temperate bird communities but not to tropical and subtropical communities.

Plotting total Deep Canyon bird density versus season reveals patterns of change that correlate with life zones (fig. 12). Lower Sonoran habitats (except valley floor) contain the most birds in spring. In the Upper Sonoran habitats, bird numbers are highest in fall and winter, whereas the Transition Life Zone contains the most birds in spring and summer. These seasonal changes in bird density reflect seasonal changes in weather and food availability. Thus, valley floor bird density is highest in winter, when the weather is mild and annual plants are abundant, and lowest in fall when the desert is hot and dry. Bird numbers in the coniferous forest, in contrast, are highest in summer, when daytime temperatures are mild, and lowest in winter when the forest is cold and snowy.

Mean annual bird density (fig. 12) shows a fairly good correlation with plant foliage volume. The chaparral, however, is an exception to this trend. Relative plant cover is greatest there, but resident bird density is lowest (see table 6). Even the valley floor contains more birds than the chaparral. Why? Perhaps the thick chaparral vegetation adversely affected our ability to detect birds. If so, this could account for the low chaparral density. I do not believe this is the case, however, for two reasons. First, chaparral bird density is lowest in the spring when singing males are conspicuous and easily detectable. Second, the piñon-juniper woodland vegetation is nearly as thick as the chaparral but has a much higher bird density. Thus, some factor(s) other than low detectability presumably contribute(s) to the chaparral's low bird density. Interestingly, the chaparral bird community is the most diverse of all Deep Canyon's communities, despite low bird numbers.

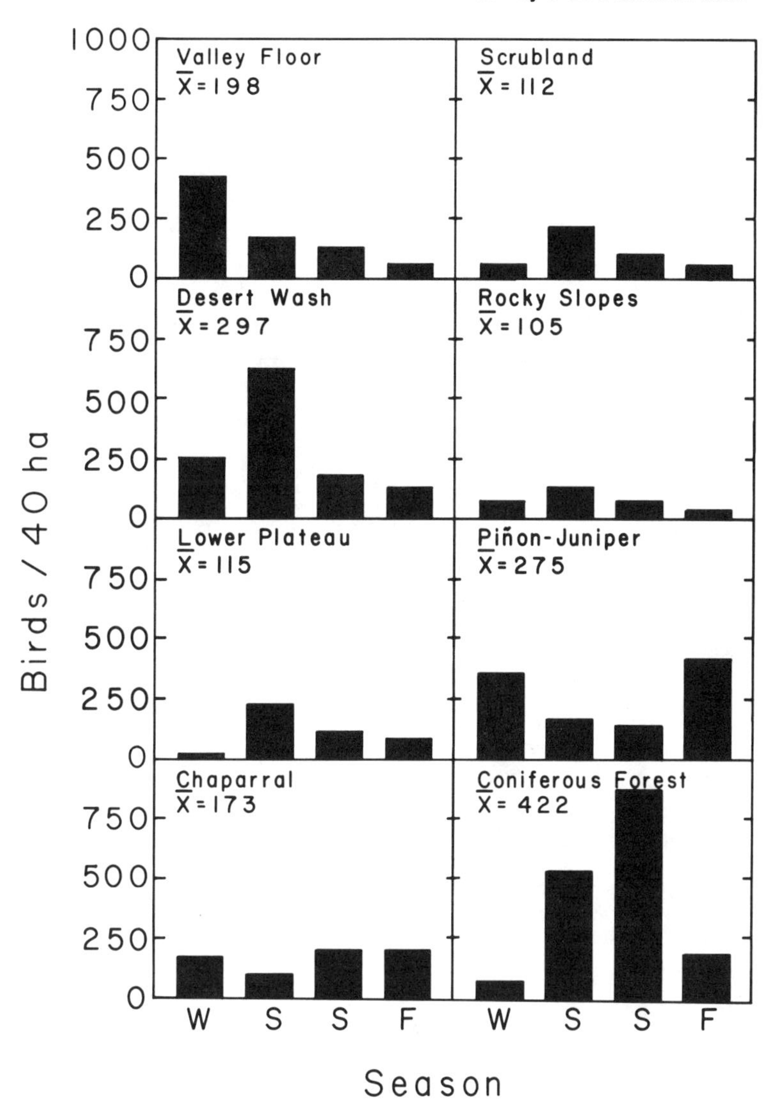

FIGURE 12 Seasonal change in total bird density and mean annual bird density (X̄) in eight different Deep Canyon habitats.

Species Diversity

The number of species in a community is the first and oldest concept of diversity and is called species richness. Determining species richness in a community of perennial plants is easy because plants are immobile. Bird species, however, constantly enter and leave a given area, and determining their number is more difficult. Partly for this reason, most studies of bird species diversity consider only the breeding season (but see Rotenberry et al. 1979). Data from Deep Canyon reveal that species richness can change greatly with season and that the breeding season is not necessarily the period of maximum richness (table 4).

By combining the resident species number (table 4) with other seasonal categories, we obtain the seasonal change in species richness. For example, the coniferous forest has the lowest number of resident species (12) but the highest number of summer visitors (20). Consequently, breeding species richness (32 species) is highest in the coniferous forest. Thus, quite different pictures of community structure are possible, depending on the season studied.

Deep Canyon's habitats differ markedly in total species richness (table 4). Desert washes have the most species (55), and the lower plateau the fewest (20). The proportion of species that are residents varies from a high of 75 percent in the lower plateau to a low of 35 percent in the coniferous forest. Presumably, the forest's harsh winter limits the number of species that can live there permanently. The habitats also differ in the contribution of each seasonal category to overall structure. Migrant birds are most important in the desert wash and rocky slopes habitats, resident birds in the lower plateau, winter visitors in the valley floor, and summer visitors in the coniferous forest.

Although three or four censuses are adequate to determine the number of resident species (Anderson and Ohmart 1977), the number required to sample adequately the total community is unknown but presumably large. For example, the 55 species encountered during 142 desert wash censuses represents only 44 percent of all species known to occur there (Appendix I). In general, species richness increases with the number of censuses. Thus, comparing Deep Canyon's communities is potentially complicated by unequal numbers of censuses (see table 2). But because no significant correlation exists between number of species found and number of censuses, the patterns observed in table 4 are considered reasonably accurate.

TABLE 4 Number of Species Recorded During Year-round Censuses at Deep Canyon

Habitat	Total no. species	Resident	Summer visitor	Winter visitor	Spring migrant
Valley floor	41	18(0.44)[a]	5(0.12)	14(0.34)	4(0.10)
Scrubland	40	18(0.45)	4(0.10)	7(0.18)	11(0.28)
Desert wash	55	21(0.40)	8(0.15)	10(0.19)	16(0.31)
Rocky slopes	46	19(0.41)	4(0.09)	3(0.07)	20(0.43)
Lower plateau	20	15(0.75)	1(0.05)	0(0.00)	4(0.20)
Piñon-juniper	40	26(0.65)	3(0.08)	9(0.23)	2(0.05)
Chaparral	43	26(0.60)	10(0.23)	6(0.14)	1(0.02)
Coniferous forest	34	12(0.35)	20(0.59)	0(0.00)	2(0.06)

[a]Number of species (proportion of total).

TABLE 5 Nest Density at Deep Canyon During 1979 Breeding Season

Habitat	Hectares sampled	Nests/ha
Valley floor	4.0	0
Alluvial plain		
Scrubland	4.0	1.0
Desert wash	6.1	4.9
Rocky slopes[a]	16.2	0.7
Lower plateau	1.0	7.0
Piñon-juniper	2.0	2.5
Coniferous forest	2.0	11.5

[a]Data from Carlson 1979b.

Nest Density

I surveyed nest density during the 1979 breeding season in grids of known area by checking every plant for nests. As expected, nest density varied widely between habitats (table 5). It was highest in the coniferous forest, where the following 23 nests were found near Stump Spring Meadow on 20 June: Pigmy Nuthatch 5, Western Bluebird 4, Violet-green Swallow 4, Dark-eyed Junco 3, Mountain Chickadee 3, Chipping Sparrow 1, Common Flicker 1, Purple Finch 1. The lower plateau ranked second with the following 7 nests found during a survey of habitat similar to that illustrated in chapter 8 (see fig. 36): House Finch 3, Cactus Wren 2, Black-throated Sparrow 1, Scott's Oriole 1. The lower plateau is very heterogeneous, however, and I found only one nest (Black-throated Sparrow) in 2 ha of uniform burrobush scrub habitat located 100 m from the above site. House Finches nested abundantly in the desert wash habitat and accounted for 37 percent of the following nests: House Finch 10, Verdin 6, Cactus Wren 6, Phainopepla 4, Black-tailed Gnatcatcher 1.

DETERMINANTS OF COMMUNITY STRUCTURE

Natural communities are enormously complex and ever-changing products of a multitude of historical, physical, and biological factors, of which the most important are: (1) the physical environment—especially climate and weather; (2) the quantity and quality of essential resources such as water, food, nest sites, and cover; and (3) interspecific interactions—especially competition and predation. The relative importance of these factors as determinants of community structure is poorly known, but in all probability it varies with community type (for reviews see Pianka 1978, Krebs 1978, Pielou 1974). Data from Deep Canyon illustrate how each of these factors operate to shape community structure.

Climate has a clear influence on community structure, with Deep Canyon's harsher habitats—valley floor and coniferous forest—having the lowest resident bird species diversity (table 6). Climate's effect on birds can be direct (i.e., physiological), through its impact on energy and water balance, and/or indirect (i.e., ecological), through its influence on vegetation and food availability. Birds are capable of a remarkable degree of physiological adjustment to differing climates, however; and except for truly extreme environments such as hot, waterless deserts or the frigid arctic, they are most often limited by climate's effect on vegetation structure and food availability (see Dawson and Bartholomew 1968). Deep Canyon's adjacent scrubland and desert wash habitats illustrate this well. They experience the same macroclimate, yet the desert wash supports much denser vegetation and consequently has both more individuals and more bird species than the scrubland (table 6, fig. 12). Indeed, in terms of annual energy flow and bird density, the desert wash more closely resembles a coniferous forest or piñon-juniper woodland than a desert.

Resource-based interspecific competition is often assumed to be one of the major driving forces behind bird community structure (e.g., Cody 1974). Many ecological studies assume food is limited, divide the resident birds into groups that forage in the same manner (guilds), and then analyze community structure on the basis of how food is partitioned (e.g., Holmes et al. 1979, Szaro and Balda 1979). Although Deep Canyon's bird communities have yet to be analyzed in this way, some evidence suggests that competition helps shape their structure. One example is springtime competition between resident and migrant insectivores.

Warblers and vireos are abundant spring migrants in the desert washes. These small insectivores arrive at the same time that Verdins and

TABLE 6 Comparison of Resident Bird Communities

Habitat	Density[a] birds/40 ha	Richness[b]	Diversity[c] (H)	Equitability ($J = H/H$ max)
Valley floor	94	18(23)	1.12	0.39
Alluvial plain				
Scrubland	103	18(23)	1.73	0.60
Desert wash	238	21(30)	1.95	0.64
Rocky slopes	97	19(23)	2.10	0.71
Lower plateau	106	15(16)	1.71	0.63
Piñon-juniper	202	26(29)	2.32	0.71
Chaparral	87	26(36)	2.58	0.79
Coniferous forest	206	12(32)	1.55	0.62

[a]Mean annual density of residents.

[b]Number of resident species (number of breeding species).

[c]Calculated from the Shannon-Weiner function $H = -\sum_{i=1}^{S} p_i \ln p_i$.

Black-tailed Gnatcatchers (small, resident insectivores) are feeding nestlings and presumably at their peak annual energy demand. Competition between resident and migrant insectivores could be keen in insect-poor years, because the estimated daily energy expenditure of the migrants is 71 percent of that of the residents (table 7; see *community energy demand* below for method of calculation). The migrant's energy demand is actually even greater than this because many (certainly half) of the unidentified birds tallied during spring censuses were warblers. If half of the unidentified bird's daily energy expenditure is added to that of the migrants, the resulting total (45.7 kJ/ha) becomes nearly twice that of residents. These rough calculations suggest that migrating warblers and vireos could have a negative impact on the resident Verdin and gnatcatcher populations. I have frequently seen Verdins stop foraging to attack nearby warblers, which also suggests that these species are treated as competitors.

Competition studies that consider only birds overlook potentially important interclass interactions. For example, granivorous birds are major components of all of Deep Canyon's bird communities. Assuming seeds are limited, there might be competition not just among granivorous birds but between all the major granivore groups—birds, rodents, and ants (e.g., Brown and Davidson 1977). Plotting mean bird and rodent densities simultaneously (fig. 13) reveals that rodents are much more numerous than birds and that birds are least common where rodents are most abundant. Could this correlation reflect interclass competition? To evaluate whether Deep Canyon's rodents, birds, and ants compete, we need information on the proportion of the available seeds harvested by each. This information is unavailable, but a rough guess is possible using Gordon's (1978) data for desert harvester ants (*Veromessor pergandei*) of Deep Canyon's alluvial plain. She found that the ants remove an average of 0.5–4.5 percent of the available seeds each month and that in winter the ants' daily energy harvest is 418 to 485 kilojoules (kJ)/ha. I estimate that granivorous birds in the same habitat in winter need 36 kJ/ha per day, less than 10 percent of the value for ants. This suggests that granivorous

TABLE 7 Average Spring Density and Daily Energy Expenditure (DEE) of Birds in the Desert Wash Habitat

	Density birds/40 ha	DEE kJ/ha
Residents[a]	37.73	27.18
Migrants[b]	20.53	19.37
Unidentified birds	60.49	52.68[c]

[a]Verdin and Black-tailed Gnatcatcher.
[b]Warblers and vireos.
[c]Calculated from Walsberg's (1980) equation for nonflight foragers using an assumed mass of 8 g.

birds harvest a minute fraction (<0.5 percent) of the available seeds and argues against severe competition between the different granivore groups. Apparently, the patterns represented in figure 13 result from factors other than competition.

Differences in Verdin and Black-tailed Gnatcatcher density at Deep Canyon suggest how predation might affect community structure. Verdins and gnatcatchers are small (6-g) insectivores with similar foraging habitats—although gnatcatchers tend to forage somewhat lower in the vegetation. During the breeding season both species produce two clutches of four eggs each, but they build quite different nests. The Verdin's nest is a hollow sphere, whereas the gnatcatcher's is a deep, open cup. Despite their many similarities, Verdins are almost twice as abundant as the gnatcatchers in Deep Canyon's desert washes (17.4 versus 10.4 birds/40 ha). I suspect this difference is partly the result of nestling predation by

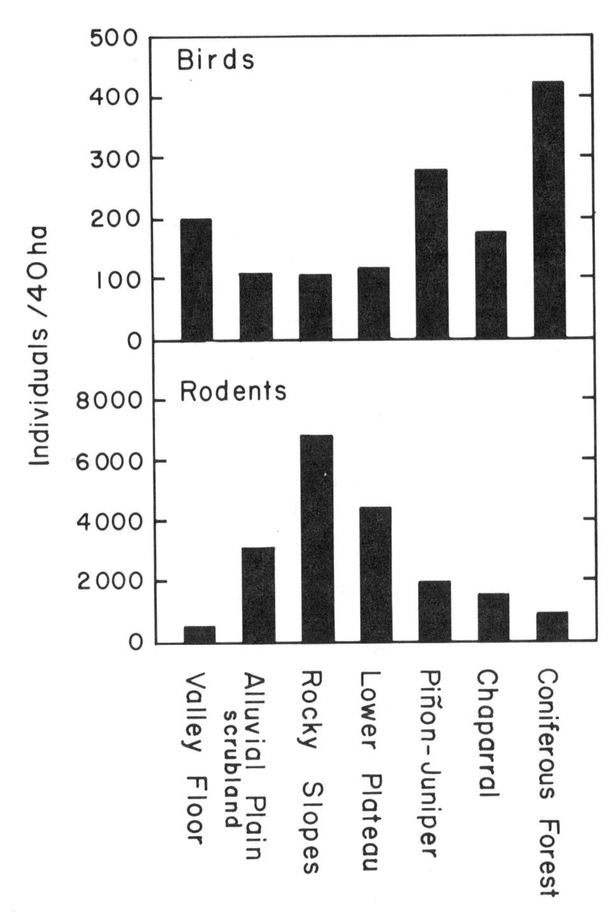

FIGURE 13 Mean annual bird density (above) and rodent density (below) throughout the Deep Canyon Transect (rodent data from Ryan 1968).

Loggerhead Shrikes. In spring, shrikes often perch atop palo verde trees and locate nests by watching the movements of adults. Once a nest is found, the shrike swoops in and nabs the nestlings. A pair of shrikes using this technique can methodically raid virtually every nest in a wash. The gnatcatcher's simple cup nest offers no obstacle to the shrikes, but the Verdin's thorny, spherical nest provides some protection. Hence, the lower gnatcatcher density at Deep Canyon may be due to shrike predation.

COMMUNITY ENERGY DEMAND

Understanding the functional position of birds in communities requires knowledge of the rate of energy flow through the component populations. Population energy expenditure cannot be measured directly, however, and must be estimated. Several estimates of energy flow through bird communities exist (e.g., Wiens and Innis 1974, Holmes and Sturges 1975, Weiner and Glowacinski 1975, Wiens and Scott 1975, Furness 1978); however, their reliability is unknown.

Existing models that estimate energy flow through bird populations vary greatly in their complexity. Some (e.g., Wiens and Innis 1974) are extremely complex, requiring input of over 20 demographic and phenological variables. Whether such complexity leads to accuracy is unknown. Indeed, the accuracy of energy expenditure estimates for free-ranging birds seldom has been measured, and errors may be large. Furness (1978) determined that his energy flow estimate of a sea bird community had a 95 percent confidence limit of ± 50 percent. Errors in this range may be common, as Weathers and Nagy (1980) found that a time-budget model of daily energy expenditure in free-ranging Phainopeplas underestimated the true energy expenditure by 40 percent. Until this field of study becomes more fully developed, it seems prudent to use simple models involving the fewest assumptions. Consequently, I estimated the average seasonal energy expenditure of bird populations at Deep Canyon in the following way. The daily energy expenditure (DEE) of an individual of average mass (values given in Appendix I) was calculated from the following equations of Walsberg (1980): for species that forage in flight (swallows, swifts, hummingbirds), kJ/day $= 13.64\, g^{0.663}$; for all others, kJ/day $= 8.96\, g^{0.653}$. The individual DEE multiplied by the estimated population density gives the population energy requirement. Because the data base underlying Walsberg's equations is very heterogeneous, this estimate should be regarded as only a first order approximation, emphasizing generality at the expense of precision.

The magnitude of the error involved in this simple estimate is unknown but may not be too large, as the following data suggest. Wiens and Innis (1974) estimated the total energy flow through a Dickcissel (*Spiza americana*) population from May through August as 11.2 megajoules (MJ)/ha. The mean population density during the 123-day period was 0.81 birds/ha. Walsberg's equation gives 77.09 kJ(bird day)$^{-1}$ as the DEE of a 27-g Dickcissel. Thus, I estimate the energy requirement of the Dickcissel population as 7.7 MJ/ha (0.81 birds/ha × 77.09 kJ/bird day × 123 days). This is only 31 percent less than the estimate derived from Wiens and Innis' much more complicated model. Since the accuracy of Wiens and Innis' model is unknown, I feel that, for community-level studies, the advantages of the simpler method of estimating energy flow far outweigh the inherent errors.

The total yearly energy flow through Deep Canyon's bird communities (estimated as above) is illustrated in figure 14. Although the piñon-juniper community has the highest energy flow, half is because of one species—the Common Raven. In chapter 9, I describe the unusual circumstances underlying the raven's dominance of the piñon-juniper com-

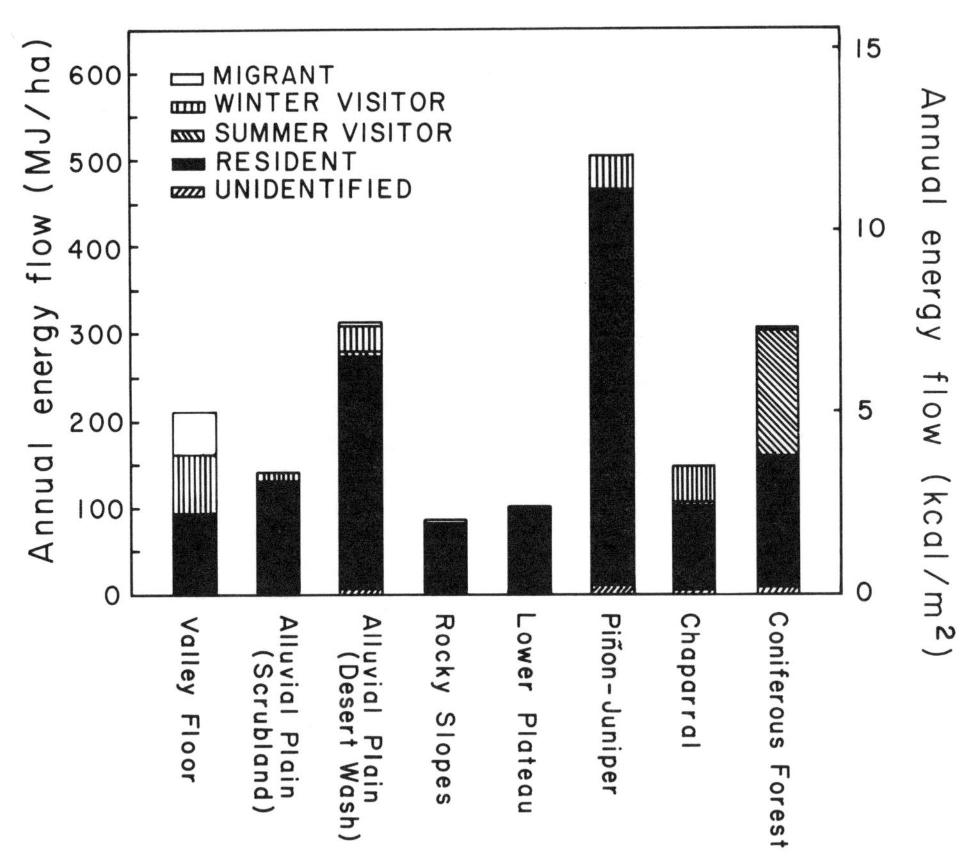

FIGURE 14 Annual energy expenditure by Deep Canyon's birds partitioned according to seasonal status. Note: half the annual energy expenditure in the piñon-juniper habitat was attributable to the Common Raven (see chapter 9 for explanation).

munity. Omitting the raven's energy expenditure, the annual piñon-juniper energy flow becomes 249 MJ/ha, slightly less than that of the desert wash community. Figure 14 reveals that the proportion of the annual energy flow attributable to resident birds varies from a high of 99 percent (lower plateau) to a low of 40.9 percent (valley floor). In most habitats, winter visitors are second in importance to residents, accounting for up to 34 percent of the annual energy flow. In the coniferous forest, summer visitors account for 47 percent of the annual energy expenditure by birds, despite the fact that they are present for only half of the year. These data show that resident species may be less significant to total community function than is sometimes assumed.

four

VALLEY FLOOR

The floor of the Coachella Valley is a parched region of shifting, wind-blown sand and sparse vegetation. Its scant rainfall, scorching summers, and fierce spring winds make life there difficult for birds. The generally flat terrain is dominated by small sand hummocks, formed around evenly spaced creosote bushes (fig. 15). The open, low plants provide little protection from snakes or the violent spring winds. Consequently, most birds nest in the dense stands of honey mesquite (*Prosopis glandulosa*) that top the larger dunes or in introduced trees such as tamarisk (*Tamarix aphylla*). Four of the valley floor's perennial plants are large, conspicuous, and abundant—creosote bush, cattle spinach (*Atriplex polycarpa*), four-winged saltbush (*A. canescens*), and honey mesquite. The barren sand between the larger plants is punctuated by herbs and the perennial grass panicum (*Panicum urvilleanum*).

Of all Deep Canyon's habitats, the valley floor has been most disturbed by man. Real estate and agricultural developments have replaced much of the original desert, with profound consequences for the region's biota. The effect of man's disturbance even extends beyond the sprawling urbanized areas, as the remaining native regions of the valley floor have been invaded by introduced species. Far from the shopping centers, in relatively pristine areas, the Starling is the second most common resident bird; and tumbleweed (*Salsola iberica*), abu-mashi (*Schismus barbatus*), and heron's bill (*Erodium cicutarium*)—all introduced weeds—blanket the dunes. Tumbleweed forms dense thickets on the formerly barren sand and attracts large flocks of finches in winter. Although the following analysis does not cover the valley's developed regions, the birds found there are listed in Appendix I under Human Habitats.

Most bird species avoid the valley floor's hot, dry summers and appear there only during winter when conditions are relatively moist and mild. Food is also more abundant then, as annual plants, which comprise about 70 percent of the flora (Zabriskie 1979), appear predominantly in winter. Winter visitors constitute a high proportion of the avifauna (fig. 16), and only four resident bird species prefer this harsh habitat, reaching

FIGURE 15 Small sand hummocks accumulate around widely spaced creosote bushes on the valley floor. Indio Hills and Little San Bernardino Mountains (beyond) in the background.

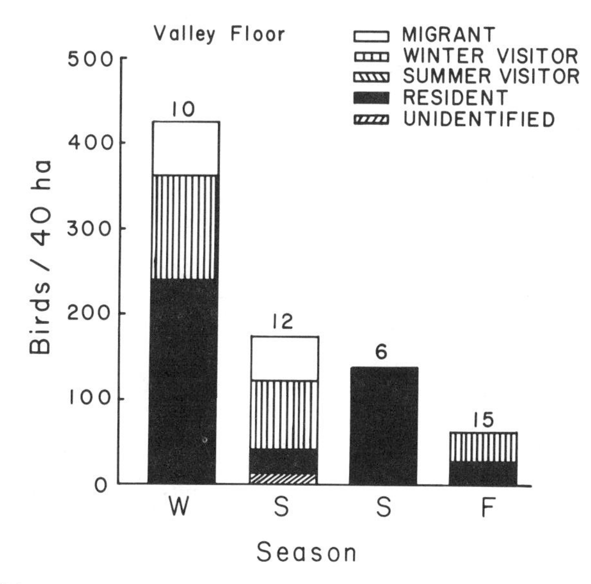

FIGURE 16 Seasonal change in bird density on the Coachella Valley floor. Density for each seasonal class (upper right) is indicated by the size of the bar. Values are means for the indicated number of strip transect censuses (numbers above bars). From left to right, bars represent: winter (W = December–February), spring (S = March–May), summer (S = June–August), and fall (F = September–November).

their highest density there. These are the Red-tailed Hawk, Burrowing Owl, Say's Phoebe, and Starling.

BIRD CENSUSES

A single 1.8-km strip transect has been established on the valley floor 3.2 km southeast of Thousand Palms village (Section 27; T. 4 S., R. 6 E., Thousand Palms Quadrangle). It crosses a broad plain (see fig. 15) containing scattered sand dunes, some of which are topped with honey mesquite.

Forty-three censuses reveal that bird density peaks during winter and declines through spring and summer to a fall low of 58 birds/40 ha (see fig. 16). Winter is both the season of greatest bird abundance and the season of greatest energy expenditure. Fifty-nine percent of the annual energy expenditure by birds occurs then (fig. 17).

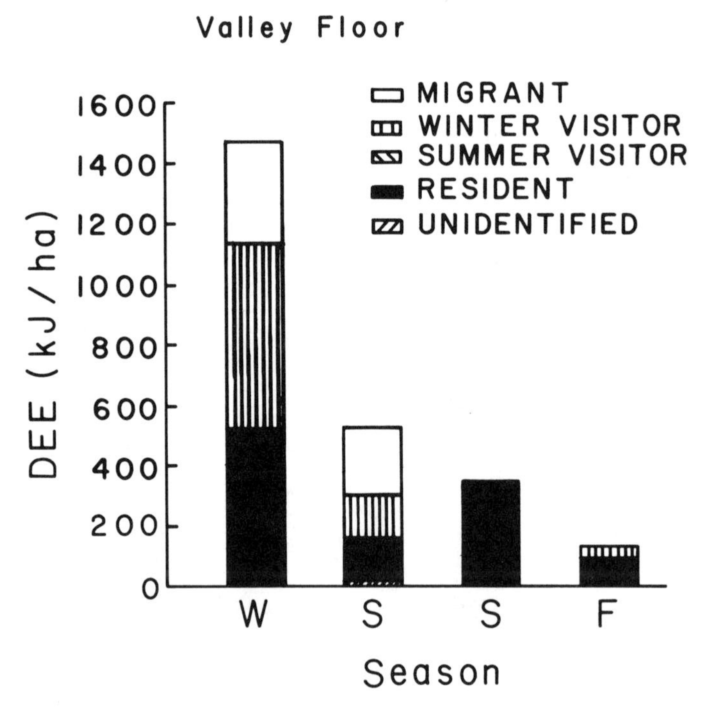

FIGURE 17 Average daily energy expenditure (DEE) by birds on the valley floor. Winter visitors account for 34 percent of the annual total, even though they are present only part of the year.

Residents

Eighteen species occur year-round on the valley floor (table 8). Most of these species breed there, but some, like the Golden Eagle and Common Raven, nest in the adjacent hills and merely forage on the valley floor.

The majority of valley floor residents are solitary and either insectivorous or raptorial—having food habits characterized by high dietary water intake. Only four species are primarily granivorous—House Finch, Mourning Dove, Gambel's Quail, and Abert's Towhee. Most niches in this desert environment are, therefore, occupied by species with moist diets. Note, however, that the four granivorous species comprise 75 percent of all individuals and account for 55 percent of the total resident energy expenditure (table 8). This is contrary to what might be expected of waterless deserts. Apparently, development of the Coachella Valley has so increased water availability that the gregarious, seed-eating species have been able to greatly increase their numbers.

The House Finch, by far the most abundant valley floor resident, accounts for 77 percent of all resident birds on an annual basis (table 8). House Finches are nomadic and gregarious (even when breeding) and, in

TABLE 8 Average Yearly Density and Daily Energy Expenditure (DEE) of Permanent Residents Recorded During 43 Censuses of the Valley Floor Habitat, 1979 and 1980

Species	% Occurrence	Density		DEE	
		birds/40 ha	% of total	kJ/ha	% of total
House Finch	79.1	70.81	77.35	112.18	47.00
Starling	16.3	5.06	5.53	19.00	7.96
Mourning Dove	18.6	3.19	3.48	16.28	6.82
Red-tailed Hawk	37.2	1.85	2.02	39.24	16.44
Loggerhead Shrike	32.6	1.74	1.90	4.75	1.99
Mockingbird	27.9	1.53	1.67	4.29	1.80
Say's Phoebe	27.9	1.34	1.46	2.19	0.92
Verdin	23.3	1.33	1.45	1.01	0.42
Roadrunner	18.6	1.03	1.13	9.56	4.01
Black-tailed Gnatcatcher	20.9	1.02	1.11	0.67	0.28
Common Raven	9.3	0.72	0.79	13.34	5.59
Gambel's Quail	9.3	0.61	0.67	3.92	1.64
Burrowing Owl	11.6	0.51	0.56	3.04	1.27
Cooper's Hawk	7.0	0.30	0.33	3.08	1.29
Total of 4 others[a]	–	0.50	0.55	6.13	2.57
TOTAL		91.54	100.00	238.68	100.00

[a]Listed in decreasing density: American Kestrel > Golden Eagle = Abert's Towhee = Black-throated Sparrow.

this respect, resemble the nomadic arid-zone granivores of southern Africa (MacLean 1974). Because of their nomadic habits, House Finches exhibit marked seasonal density changes (see chap. 13, table 44) and accordingly differ from other valley floor residents. The seasons of peak House Finch abundance are winter (215 finches/40 ha) and summer (104 finches/40 ha). In winter, House Finches are attracted to the valley floor by the dense stands of thistle and sprouting annuals. But, what attracts House Finches to the valley floor in midsummer when the weather, from a human's standpoint, is unbearably hot? The answer is uncertain, but a simultaneous influx of Mourning Doves suggests that seeds may be numerous then. If so, both the finches and doves could be using their strong powers of flight to exploit the abundant seeds by foraging during the cooler mornings and then retreating to the developed areas' shade and water at midday.

Little information exists concerning bird densities in other desert regions with which Deep Canyon's values can be compared. Franzreb (1978) found a breeding bird density of 25.3 birds/40 ha in open creosote brush scrub near California's Algodones dunes, a value nearly identical with the spring resident bird density (25.4 birds/40 ha) found in this study (fig. 16).

Winter Visitors

Winter is the season of greatest bird activity on the valley floor. Many species are attracted to the Coachella Valley by the mild winter temperatures, which are among the highest in California. Winter rains produce a luxuriant growth of annual plants, and small granivorous birds find the sprouting seeds and green seed heads ideal food. Insects thrive on the growing annuals, and they in turn attract insectivorous birds such as Cliff Swallows, Meadowlarks, and Savannah Sparrows. (The latter species reverses the usual pattern by migrating northward from Baja California to southern California in winter.) These three species, together with the House Finch, White-crowned Sparrow, Sage Sparrow, and Brewer's Sparrow, comprise 82 percent of the valley floor's birds in winter.

Ephemeral ponds form in the Coachella Valley's low-lying regions following heavy rains (fig. 18). They may persist for several weeks, attracting shorebirds and waterfowl to the valley floor, and account for the Pintail's presence on the winter visitor list (table 9). Because the Pintail is a large and abundant bird, the winter visitor category had an unexpectedly high daily energy expenditure (fig. 17). During drier, more normal years, waterfowl are not found in this part of the Coachella Valley. Thus, winter visitors are typically less significant than figures 16 and 17 indicate.

Most winter visitors reach their peak density during the months of December through February. Brewer's and Sage Sparrows, however, arrive on the valley floor in mid-October—one-and-a-half months earlier than many other winter visitors—and attain higher densities in fall than winter (5.57 versus 2.22 birds/40 ha for Brewer's Sparrow and 14.67 versus 4.43 birds/40 ha for Sage Sparrow). This indicates that some of the fall birds are southbound migrants.

FIGURE 18 A temporary pond on the Coachella Valley floor produced by heavy rains. Water remained at this site for several months following the storm (photograph courtesy of Lloyd Tevis, Jr.).

TABLE 9 Average Density and Daily Energy Expenditure (DEE) of Winter Visitors Recorded During 10 Winter Censuses (Dec.–Feb.) of the Valley Floor Habitat, 1979 and 1980

Species	% Occurrence	Density		DEE	
		birds/40 ha	% of total	kJ/ha	% of total
Western Meadowlark	20.0	53.28	43.92	238.28	39.48
Pintail	10.0	13.76	11.34	261.81	43.38
Savannah Sparrow	20.0	11.54	9.51	16.44	2.72
Phainopepla	80.0	9.71	8.00	17.33	2.87
Violet-green Swallow	20.0	6.66	5.49	13.68	2.27
White-crowned Sparrow	10.0	6.22	5.13	12.56	2.08
Yellow-rumped Warbler	60.0	5.74	4.73	6.51	1.08
Sage Sparrow	40.0	4.43	3.65	5.82	0.96
Common Flicker	20.0	3.76	3.10	22.20	3.68
Western Bluebird	10.0	2.22	1.83	4.28	0.71
Brewer's Sparrow	10.0	2.22	1.83	2.38	0.39
Blue-gray Gnatcatcher	10.0	0.88	0.73	0.60	0.10
Bewick's Wren	10.0	0.44	0.36	0.42	0.07
Sage Thrasher	10.0	0.44	0.36	1.27	0.21
TOTAL		121.30	99.98	603.58	100.00

Summer Visitors

Rigorous weather and a dearth of suitable nesting sites combine to make the valley floor an unattractive breeding habitat for summer visitors. Accordingly, only three species of summer visitors were detected during eighteen spring and summer valley floor censuses—the Ash-throated Flycatcher, Rough-winged Swallow, and Western Kingbird. Their combined density is a meager 1.95 birds/40 ha, and they are insignificant components of this habitat's avifauna (figs. 16 and 17).

Migrants

Terrestrial species migrating through the Coachella Valley include swallows, warblers, sparrows, flycatchers, and raptors. Winter and spring are the seasons of peak migration, as most small songbirds bypass the Coachella Valley during fall migration, traveling south along the mountain peaks instead. Migrating waterfowl and shorebirds occur on the valley floor during both seasons, but none were detected during censuses.

Of the six species of migrants found during spring censuses, two, the Cliff Swallow and Barn Swallow, account for 89 percent of all individuals (table 10). Cliff Swallows are early migrants, appearing on the valley floor in late February. They account for most of the late winter migrants represented in figure 16.

As a group, migrants account for 14.7 percent of the birds found on the valley floor and use 24 percent of the yearly energy that flows through valley floor birds. This disproportionately high energy flow reflects the effect of a single, large migrant—the Turkey Vulture—on energy estimates. Migrating Turkey Vultures were never numerous (maximum density 6.7 birds/40 ha), but because they are very large they account for 47 percent of the total annual energy expenditure by migrant birds. Because it feeds mainly on carrion, the Turkey Vulture exerts minimal competitive pressure on other birds, despite its high energy demand.

TABLE 10 Average Density and Daily Energy Expenditure (DEE) of Spring Migrants Recorded During 12 Spring Censuses (Mar.–May) of the Valley Floor Habitat, 1979 and 1980

Species	% Occurrence	Density		DEE	
		birds/40 ha	% of total	kJ/ha	% of total
Cliff Swallow	25.0	23.67	45.40	60.75	28.51
Tree Swallow	16.7	22.57	43.29	52.30	24.54
Turkey Vulture	16.7	2.96	5.68	81.33	38.16
Barn Swallow	8.3	1.48	2.84	3.55	1.67
Vaux's Swift	8.3	0.73	1.40	1.69	0.79
Swainson's Hawk	16.7	0.73	1.40	13.49	6.33
TOTAL		52.14	100.01	213.11	100.00

five

HUMAN HABITATS

". . . all man's progress has been made at the expense of damage to his environment which he cannot repair and could not foresee."

—C. D. Darlington.
Regius Professor of Biology, Oxford.

The original Spanish name for the Coachella Valley is "*La Palma de la Mano de Díos*"—the palm of God's hand (Jaeger 1957). One's expectation that people would be inclined to settle in God's palm is borne out by the large expanses of urban and agricultural areas (human habitats) now found in the Coachella Valley. The kind of development that has occurred reflects man's physiology as much as his economic and societal traits. For unlike many animals, man needs drinking water to survive in the desert. Thus, development of the desert has increased the available water and thereby changed the balance of life. As the valley floor changed from former lake bed and windblown dunes to agricultural fields and suburbs, one community of birds replaced another. Areas that formerly supported cacti, smoke trees (*Dalea spinosa*), palo verde, and their associated birds—the Verdin, Cactus Wren, Black-tailed Gnatcatcher, and Phainopepla—are now occupied by buildings and golf courses (fig. 19). Ducks and grebes paddle about where Roadrunners formerly stalked lizards. Unfortunately, because we are painfully ignorant of what the desert was originally like, we cannot appreciate what has been lost.

The only bird apparently extirpated from the region because of man is the California Condor, *Gymnogyps californianus* (see chap. 13). Formerly abundant throughout California, condors have been poisoned, trapped, and shot to the brink of extinction. At present no more than twenty to thirty condors survive in the wild.

Valley development probably reduces other native species' populations in direct proportion to the amount of habitat lost. Thus, when 50 percent of Deep Canyon's alluvial plain has been developed, roughly half of the birds native to that habitat will have perished. This "rule" cannot be applied strictly, however, as some species require a minimum habitat size. Furthermore, extinction may occur even if large amounts of habitat exist, for the number of birds may fall below the threshold required to stimulate reproduction. This effect may have contributed to the extinction of the Passenger Pigeon (*Ectopistes migratorius*), a former resident of eastern North America.

FIGURE 19 Development is altering the desert's balance of nature. Golf course water hazards, such as this one at the Ironwood Country Club, attract birds that otherwise would not occur at Deep Canyon.

Although examples of major habitat types should be preserved to protect their unique gene pools, in most cases we can only guess how much habitat to set aside. Lack of adequate information about the population dynamics of even the most common species is a major stumbling block to sound management and conservation. How much more development can occur in the Coachella Valley before rare and little-known species like the Crissal Thrasher vanish? Apparently we have decided to find the answer empirically.

Although conservation of dwindling natural resources is vital to our self-interest, some attempts to preserve the biota of "God's palm" are philosophically and biologically unsound. One example is ordinances that prohibit collecting reptiles within city limits. Children noosing lizards for pets, or even commercial collectors, are no match for the destructive power of bulldozers. Development, not collecting, will cause the demise of the reptiles. Passing an anticollecting ordinance, while doing nothing to limit development, is like giving aspirin to a man dying from infection: it creates the illusion that something useful is being done but does not stem the disease. As urbanization flows rapidly up the alluvial plain, the chance to appreciate and study this desert's unique life fades away forever.

To understand man's effect on Deep Canyon's birds, we need information that predates the valley's development. The oldest systematic bird study of the region is that of Grinnell and Swarth (1913). Although their study contains little information on the birds of Deep Canyon's lower elevations, it provides a basis for identifying the general changes caused by development. The increased availability of water, notably golf course water hazards, has had a particularly marked effect. Ducks, grebes, loons, and shorebirds that formerly passed over Deep Canyon on migration now stop. These water-loving birds make up most of the thirty-four bird species known only from human habitats (Appendix I). Five species, three of which are introduced exotics, now regularly breed on the transect but apparently did not do so in 1908. These are the Mallard, Rock Dove, Spotted Dove, Killdeer, and Starling.

Thirteen native resident species have probably benefited from the valley's development because of increases in food, water, and nest sites. These are Red-tailed Hawk, Common Raven, American Kestrel, Mockingbird, Gambel's Quail, American Robin, Mourning Dove, Western Meadowlark, Ground Dove, Brewer's Blackbird, Barn Owl, House Finch, and Screech Owl. Of these, the greatest beneficiary has been the House Finch, now Deep Canyon's most abundant species.

six

ALLUVIAL PLAIN

The Santa Rosa Mountains' soil is poorly developed, its hillsides steep, and its ground cover sparse. These features combined contribute to the rapid runoff of stormwaters, which coalesce as they descend the slopes, carving steep-walled gorges, and carrying debris down-mountain to the Coachella Valley. Downpours of enormous magnitude, especially in earlier epochs, deposited the sediments that today form Deep Canyon's alluvial plain (fig. 20). As floodwaters reach the base of the mountains and spread out across the plain, their carrying power diminishes. Consequently, large boulders and rocks accumulate near the mountains' base. Farther down, the plain becomes progressively smoother, with rocky ground giving way to sand. Over a 5 km distance the alluvial plain's elevation decreases from 365 m near the mountain to 120 m at its point of coalescence with the alluvial deposits from Carrizo and Dead Indian Canyons. From there the plain forms a large, smooth bajada that gradually merges with the windblown sand of the valley floor. Recent stormwaters have bisected the alluvial plain, forming sandy desert washes that are separated by intervening regions of desert scrubland. The scrubland and washes are distinct habitats and, therefore, are discussed separately.

DESERT WASHES

Stormwaters from Deep, Sheep, and Coyote canyons flow across Deep Canyon's alluvial plain and create sandy desert washes. High-banked and narrow near the canyon entrances, the washes gradually broaden and branch out as they descend the plain until the distinction between them and the surrounding scrubland disappears. Large floods may fill the washes from bank to bank and create changes in the location of the watercourses, isolating remnants of the plain as elevated islands.

Plant distribution and abundance in the washes are determined by the amount of runoff received. Runoff is greatest in Deep Canyon, but because the powerful flood waters tear away plants, the vegetation is

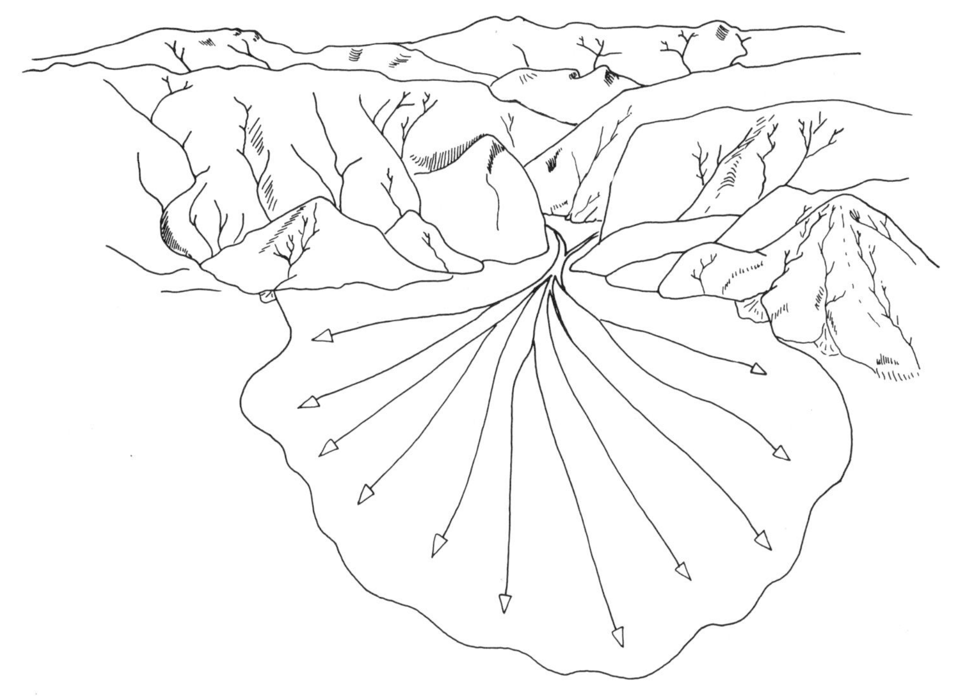

FIGURE 20 Storm waters erode the Santa Rosa Mountains, carrying debris out of Deep Canyon and depositing it on the alluvial plain.

actually less dense there than in Sheep and Coyote washes. Deep Canyon Wash is boulder strewn and barren for the first few kilometers beyond its entrance onto the alluvial plain. In contrast, Sheep and Coyote washes abruptly develop into sandy floors that support full-sized trees.

Flora

Desert washes are the most luxuriant of Deep Canyon's native desert habitats—verdant islands in a parched sea of sand and rock. The trees and large shrubs of the washes are attractive foraging and nesting sites for birds. For this reason, washes support a much higher density of birds than the other desert habitats. The washes are dominated by palo verde, a tree with green trunk and branches that may reach a height of seven meters (fig. 21). Each March the winter-deciduous palo verde produces bright, yellow flowers that last until June. The flowers, which attract numerous insects, are carefully searched by small insectivorous birds such as the Verdin. Leaf expansion occurs in May following flowering and coincides with pod development. (The palo verde is a member of the pea family, and its fruits are easily recognizable as typical pea pods.) Thus, a pulse of new, green foliage occurs in the washes at a time when the rest of the alluvial plain's vegetation is losing its leaves in response to the increasingly hot, dry weather. The palo verde's leaves persist throughout summer.

FIGURE 21 Dry, sandy desert washes bisect Deep Canyon's alluvial plain. The washes' numerous palo verde trees (right foreground), favorite nesting sites of many desert birds, attract several warbler species during spring.

This helps maintain insect numbers and thus provides food for many birds raising their second brood of nestlings. In addition, the dense, spiny palo verde branches provide abundant nest sites. Without this remarkable tree the desert washes would support far fewer birds.

Smoke tree and desert willow (*Chilopsis linearis*) are the other major trees of the wash. Like the palo verde, the smoke tree has a layer of photosynthetic tissue in its stems, but an overlying coating of minute hairs gives the stems a smokey-blue rather than a greenish cast. These hairs may discourage insects from devouring the photosynthetic layer and result in few insect larvae being available. Consequently, few birds forage within the smoke trees, but the dense, spiny branches are highly favored as nesting sites by Verdins, House Finches, and Black-tailed Gnatcatchers.

The desert willow (a member of the trumpet vine family—Bignoniaceae—rather than a true willow) is the one nonleguminous tree of the wash. It is winter-deciduous and in spring produces new leaves and large, two-lipped, lavender flowers that persist throughout summer. In late June, after most plants have set fruit, the desert willow's flowers constitute the major nectar source for Costa's Hummingbirds. At this time, hummingbirds abandon Sheep and Coyote washes for the dense stands of desert willow in Deep Canyon Wash.

Two trees, cat's claw (*Acacia greggii*) and honey mesquite, occur in the washes as low, dense, spreading shrubs. Like the palo verde, they are winter-deciduous members of the pea family. They produce new leaves and flowers in April. The cat's claw, which takes its name from the shape of its sharp, clinging thorns, was known to early desert travelers as "tear-blanket" or "wait-a-minute" (Jaeger 1965). Verdins frequently nest in cat's claw and may be protected from predators by the plant's thorns.

Six species of shrubs and subshrubs occur commonly in the washes (Zabriskie 1979). Of these, the two most abundant, chuparosa (*Beloperone californica*) and desert lavender (*Hyptis emoryi*), are of obvious importance to birds. Chuparosa, a densely compact shrub with a round crown and scarlet, tubular flowers, is drought-deciduous and loses its leaves rapidly. Desert lavender has small leaves, is slowly drought-deciduous and, except during the driest times of the year, has minute, woolly, blue-gray flowers. Both species exhibit lax seasonality, flowering at any time of year following sufficient rain. The flowers of chuparosa (Spanish for "sucking rose") are visited by many species of birds. Long-billed species, like hummingbirds and orioles, insert their bills into the flowers and use their tongues to extract nectar. Warblers, whose bills are much too short to obtain nectar in this way, pierce the corolla at its base. House Finches and Gambel's Quail bite off the corolla tubes to get at the flower's ovaries and nectar glands.

Costa's Hummingbirds frequently nest in desert lavender, binding the flower's woolly sepals together with spiderwebs to make an extremely well-camouflaged cup. They also obtain nectar from the minute flowers. Other species nesting among the desert lavender's wandlike branches include the Black-tailed Gnatcatcher, House Finch, and Verdin.

Some desert wash plants are governed by photoperiod and hence have predictable annual cycles. Others respond to the erratic rains and are thus unpredictable. The overall pattern is one of considerable year-to-year variation. Indeed, variability is such a central feature of deserts that "typical" years do not exist. Birds reflect the annual plant variations. Thus, it is impossible to predict with confidence what you will see in the desert at a given time of year. But, an "average year" can be described from the results of the present study, and this will help illustrate the underlying pattern of seasonal change.

DESERT SEASONS

Plotting air temperature and bird density in the desert wash over two cycles creates a sense of annual cycling (fig. 22). Note, however, that temperate-zone birds respond much more strongly to photoperiod than temperature, and I do not mean to imply that the change in bird numbers is a result of temperature. I use temperature merely as a convenient reference.

Bird density shows a pronounced annual cycle. It increases throughout winter, peaks in April well before the peak in air temperature, and then declines precipitously as the desert parches under the summer's sun. By fall, density has reached its seasonal low. Although this seasonal den-

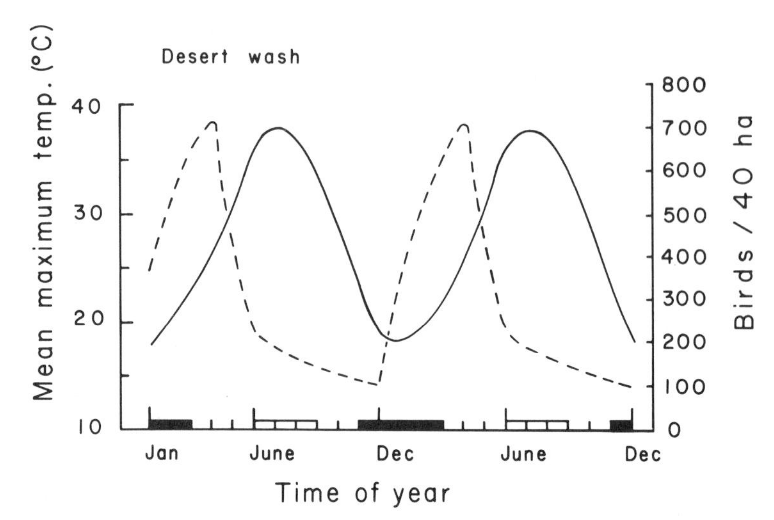

FIGURE 22 Desert wash bird density (dashed line) increases rapidly during winter when the temperature (solid line) is low. Density reaches a peak in April well ahead of the peak in temperature. The transition between the generally low temperatures of winter (black bars) and high summer temperatures (white bars) occurs quickly on the Colorado Desert.

sity change results from a multitude of interacting factors, its essential character can be revealed by describing an annual cycle.

Fall (September–November)

A friend once remarked that autumn was his favorite desert season because "everything is dead. No pollen, no birds, no plants; calm, hot, and dead. That's the way a desert is supposed to be." The description is apt, as an autumn walk down the desert wash reveals few birds—even House Finches and Gambel's Quail are scarce. The contrast between the wash's stillness in fall and its throbbing vitality in spring could hardly be greater.

By late September, the nesting season is over and fall migration has begun. Scott's Orioles and Ash-throated Flycatchers depart for their southern wintering grounds just as wintering White-crowned, Sage, and Brewer's Sparrows begin to arrive from the north. Although a few wintering Bewick's Wrens, Rock Wrens, Yellow-rumped Warblers, and Orange-crowned Warblers appear on the alluvial plain at this time, most are south-bound migrants. Bird numbers do not increase appreciably until after the first winter storms in November. Ruby-crowned Kinglets and mixed-species flocks of finches arrive on the alluvial plain in mid-October. Most, however, are migrants that pause only briefly on their way south.

In early November, Phainopeplas arrive in the desert washes, sound their single low *wurp* note from the tops of the palo verde trees, and thereby herald the approach of winter.

Winter (December–February)

Bird density in the desert washes increases three-fold during winter because of an influx of winter visitors and residents (fig. 22). In winter the most common species are the House Finch, Costa's Hummingbird, and Phainopepla. Together they account for half of all birds. The large number of Costa's Hummingbirds illustrates the importance of Colorado Desert washes as wintering areas for this species, which is a summer visitor elsewhere in California. Hummingbirds are attracted to the washes by the large concentrations of chuparosa and desert lavender. Both plants produce flowers and leaves after autumnal or winter rains. On 1 January 1977 half of all chuparosa in the desert washes had flowers. This abundant nectar source, supplemented with flying insects during periods of warm weather, made the washes ideal hummingbird habitat.

Mixed-species flocks become increasingly common in the washes and surrounding portions of the alluvial plain from late December through February. These flocks, which typically contain White-crowned, Brewer's, and Black-throated Sparrows, are occasionally joined by Cactus Wrens. Cody (1971) describes winter flock behavior in the Mojave Desert and lists twenty-five species known to join such flocks.

By the end of January, the increasing day-length initiates breeding in some individuals, and the first signs of nesting—song and social display—appear throughout the alluvial plain. A dawn chorus of House Finches, Say's Phoebes, Cactus Wrens, Gambel's Quail, and Rock Wrens becomes more vigorous with each passing day. But despite the springlike tempo, many individuals remain reproductively inactive. Some House Finches remain in flocks on the valley floor, while others move onto the alluvial plain and set up territories. The sexually slow starters are probably first year birds.

By the first week of February, Say's Phoebes, Red-tailed Hawks, House Finches, Black-tailed Gnatcatchers, and Phainopeplas have all begun to set up breeding territories. Annual plants and grasses are frequently abundant by mid-February, and on warm days flying insects appear. Springlike conditions continue throughout February, and the increasing greenness is indeed impressive for this time of year.

Spring (March–May)

Bird numbers on the alluvial plain peak between late March and early April, the height of the annual cycle (fig. 22). By the vernal equinox, most resident species are incubating eggs or feeding nestlings. Flowers and insects are abundant, and winter visitors such as White-crowned Sparrows, Brewer's Sparrows, and Bewick's Wrens have begun to sing in preparation for their northward departure. Nearly every blossom-packed palo verde tree contains migrating warblers, tanagers, orioles, and flycatchers. In some years, hoards of migrating *Selasphorus* hummingbirds invade the washes, swamping the resident Costa's Hummingbirds, which had been defending individual chuparosa bushes.

Bird numbers decline noticeably by the end of April as the winter visitors depart. Migrating warblers, however, are just reaching their peak. By the first week of May, most of the residents have fledged young, including: Black-tailed Gnatcatchers, House Finches, Roadrunners, Loggerhead Shrikes, Gambel's Quail, Killdeer, Mourning Doves, Black-throated Sparrows, Verdins, Cactus Wrens, Costa's Hummingbirds, and Say's Phoebes.

Summer (June–August)

Summer is a time of comparative dormancy. Many species leave the desert, and those that remain become less active. The weather is hot, with air temperatures near 35°C at dawn, rising to 45°C, or higher, by midday. The calm, dry dawn is no longer punctuated by the calls of Say's Phoebes, Cactus Wrens, Black-throated Sparrows, or Scott's Orioles. Instead, the monotonous chirps of House Finches and the incessant peeping of young Gambel's quail become the only regular sounds.

As the desert heats up, birds must cool themselves by panting. But because evaporative cooling entails the loss of vital body water, birds behaviorally reduce their reliance on panting by confining their activity to the early morning and retreating to the shade at midday. The shade of the breezeway at the Boyd Research Center attracts many species, and it is common to find several Cactus Wrens, House Finches, and Say's Phoebes resting there at noon. Elsewhere, a few Black-throated Sparrows may even retreat down mammal burrows to escape the sun (Austin and Smith 1974). A person hiking across the alluvial plain at midday, when the temperature is 46°C (115°F), will find the experience eerie. Amid the shimmering heat waves, an occasional desert iguana (*Dipsosaurus dorsalis*) appears, but there are no birds and no sounds, except for the wind.

The smoke trees blossom in June, providing a welcome splash of color that coincides with a bloom in flying insects. Although flying insect biomass usually peaks in summer, the year-to-year variation is great (fig. 23). During August 1979 tiny, flying gnats reached such plague proportions that censusing the washes became an agonizing chore. But man's pests are the birds' delight, as insects constitute an important source of dietary water for many species. By eating insects, the Verdin, Black-tailed Gnatcatcher, and Black-throated Sparrow can tolerate the summer's heat without drinking water, while seed-eaters like Gambel's Quail, Mourning Dove, and House Finch must drink daily.

DESERT WASH CENSUS RESULTS

The desert wash was the most thoroughly censused of all the habitats. Six transects, ranging in length from 0.62 to 1.93 km (mean 1.10 km), were established in Rubble, Sheep, and Coyote washes (Sections 4, 9, 10, 16; T.

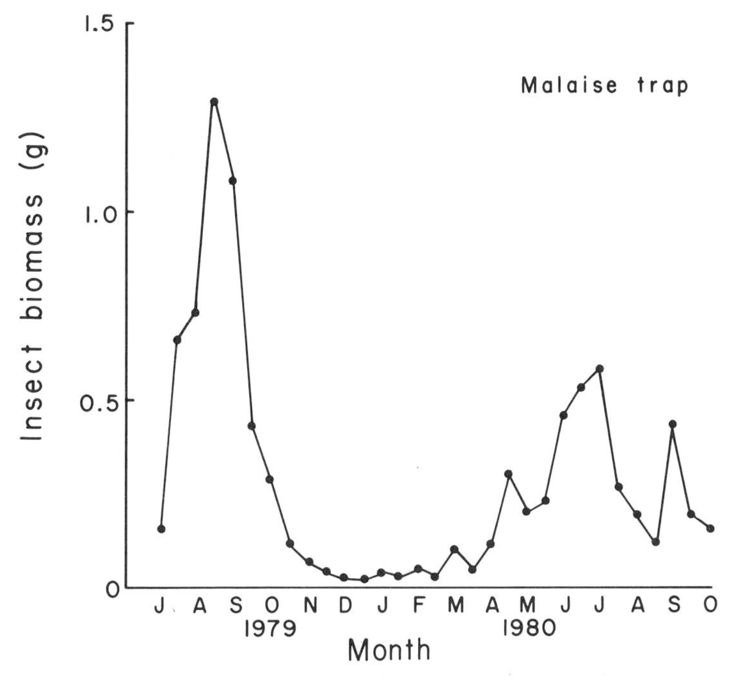

FIGURE 23 Flying insects on the alluvial plain are usually most abundant during summer, but there is considerable year-to-year variation (data courtesy of Philip L. Boyd Deep Canyon Desert Research Center).

6 S., R. 6 E., Palm Desert Quadrangle). Most of the 142 censuses were conducted between February 1979 and January 1981.

Because of the large amount of available data, bird density can be expressed on a monthly as well as a seasonal basis. Plotting bird density by month (fig. 24) shows clearly the annual cycle in numbers and illustrates the variation in density found between censuses. Density reaches its peak in April, when conditions are lush, and its nadir in October, when the washes are utterly parched from the summer's heat. Plotting bird density by season (fig. 25) reveals that residents are the dominant group at all seasons. In spring, some individuals of resident species are attracted to the washes by the abundance of food and nest sites, and this accounts for the spring peak in resident density.

Residents

Desert wash residents are listed by decreasing density in table 11. All of the residents listed nest in the desert washes, except the White-throated Swift, Prairie Falcon, and Common Raven. These three nest in the nearby canyons and cliffs and visit the washes to search for prey.

The mean annual densities given in table 11 conceal substantial seasonal density changes in the seven most abundant residents (table 12). The Mourning Dove and House Finch show large seasonal cycles, their

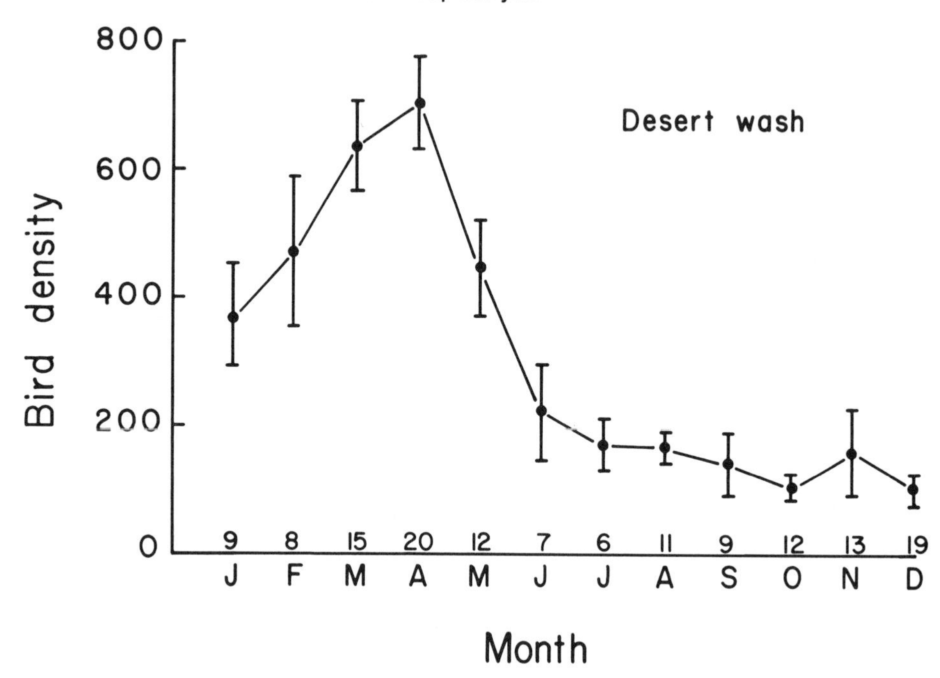

FIGURE 24 Annual cycle in bird density (birds/40 ha) in the desert wash habitat, 1979 through 1980. Points are means (±SE) for the indicated number of censuses.

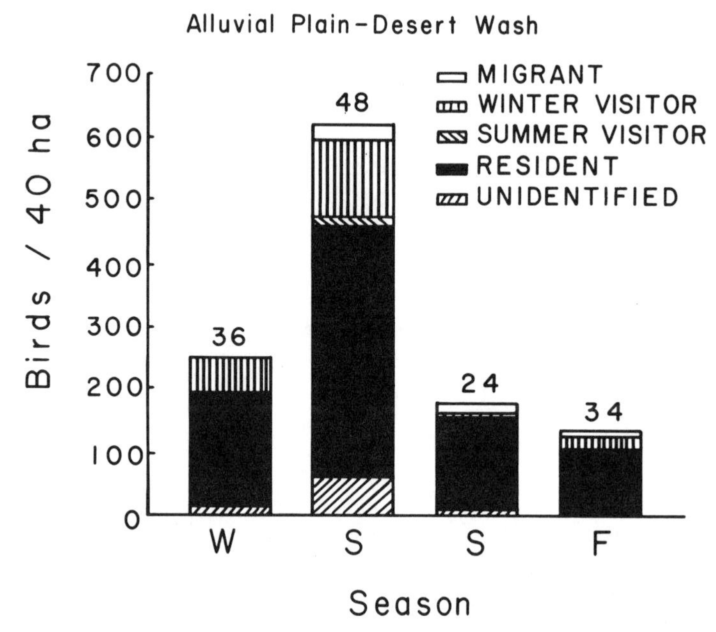

FIGURE 25 Seasonal change in desert wash bird density at Deep Canyon. See fig. 16 for explanation of symbols.

TABLE 11 Average Yearly Density and Daily Energy Expenditure (DEE) of Permanent Residents Recorded During 142 Censuses of the Desert Wash Habitat, 1979 and 1980

Species	% Occurrence	Density		DEE	
		birds/40 ha	% of total	kJ/ha	% of total
House Finch	56.4	79.02	33.26	125.19	17.20
Gambel's Quail	62.4	54.57	22.97	351.03	48.23
Mourning Dove	48.2	26.88	11.31	137.21	18.85
Costa's Hummingbird	71.6	22.17	9.33	16.35	2.25
Verdin	70.9	17.44	7.34	13.26	1.82
Black-tailed Gnatcatcher	57.5	10.36	4.36	6.81	0.94
Cactus Wren	47.5	8.37	3.52	20.85	2.86
Mockingbird	22.7	7.40	3.11	20.76	2.85
Black-throated Sparrow	24.1	5.13	2.16	6.01	0.83
Loggerhead Shrike	32.6	3.01	1.27	8.21	1.13
Say's Phoebe	9.2	0.93	0.39	1.52	0.21
Cooper's Hawk	6.4	0.47	0.20	4.83	0.66
Total of 9 others[a]		1.85	0.78	15.81	2.17
TOTAL		237.60	100.00	727.84	100.00

[a]Listed in decreasing density: Roadrunner > Abert's Towhee > Red-tailed Hawk > Brown-headed Cowbird > White-throated Swift > Starling > Prairie Falcon = Common Raven > Ground Dove.

TABLE 12 Seasonal Change in Density of Residents in the Desert Wash Habitat

	Density (birds/40 ha)			
	Winter	Spring	Summer	Fall
House Finch	57.9	183.6	8.3	3.6
Gambel's Quail	22.4	70.0	70.3	55.7
Mourning Dove	16.8	58.5	16.1	0.5
Costa's Hummingbird	34.5	28.7	2.0	14.6
Verdin	17.6	23.1	16.6	9.9
Black-tailed Gnatcatcher	10.8	14.6	8.1	5.5
Cactus Wren	5.6	5.5	22.6	5.6

numbers declining markedly in summer and fall. Both are strong fliers and leave the desert washes in summer for other nearby habitats; House Finches depart for the valley floor, whereas many Mourning Doves move to the piñon-juniper habitat (fig. 26). Gambel's Quail, much weaker fliers than either House Finches or Mourning Doves, remain on the alluvial plain year-round. Hence, the decrease in quail numbers in fall and winter probably represents natural mortality. In summer most Costa's Hummingbirds leave the washes for the chaparral of California's Coast Range

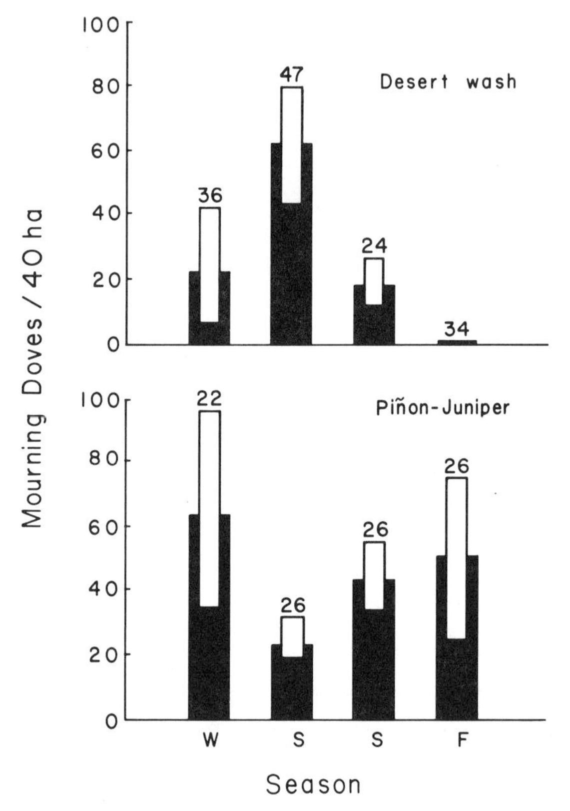

FIGURE 26 Mourning Doves prefer the desert washes in spring but the piñon-juniper woodland at other times of year. The solid bar is the mean density for the indicated number of censuses. White bars represent ±1 SE. The variance is highest during fall and winter, when the doves occur mainly in flocks.

where they may breed again. Of those remaining behind, most leave Rubble, Sheep, and Coyote washes for the desert willow blossoms in Deep Canyon Wash. The reasons for the seasonal density changes of other species are less clear.

Because residents are abundant and generally large, they dominate energy flow in the desert wash community and account for 85 percent of the annual energy flow (fig. 27). Among residents, three seed-eating species account for 67 percent of all individuals (table 11). Two of the three, Gambel's Quail and the Mourning Dove, are large. Consequently, the three seed-eating species combined account for 84 percent of the annual energy expenditure by wash residents. Gambel's Quail, the second most abundant resident, is the most important species in terms of energy flow, accounting for 41 percent of the annual energy expenditure by wash birds and 48 percent of the expenditure by residents.

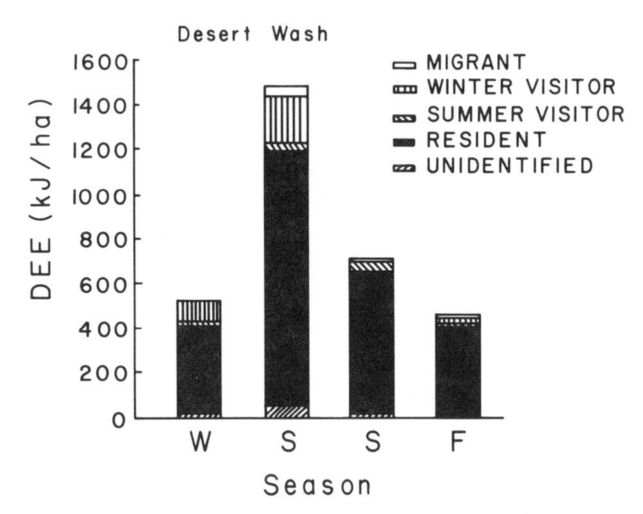

FIGURE 27 Average daily energy expenditure (DEE) by desert wash birds is highest in spring when bird density is highest. Resident species account for 85 percent of the annual energy flow.

Winter Visitors

Winter visitors are second in importance to residents in terms of both density and energy flow (figs. 25 and 27). The Phainopepla, by far the most abundant winter visitor, accounts for over half the bird density and energy flow in this seasonal class (table 13). It is even more important, in terms of energy flow, than table 13 indicates because, unlike the other winter visitors, it breeds during its stay. The peculiar breeding biology of this bird has been studied in detail by Walsberg (1977) and is reviewed in chapter 13.

Summer Visitors

Eight species occur as summer visitors in the desert washes (table 14). All these birds breed in the washes except the swallows, which nest nearby and merely forage in the wash. None of the eight is abundant, and thus summer visitors contribute little to the total wash avifauna (figs 24 and 27), accounting for less than 2 percent of the annual energy flow.

Migrants

Migrants contribute more to the desert wash's species richness than to its energy flow, with 16 migrant species encountered during spring wash censuses (table 15) and another 7 species recorded at other times during the spring (Appendix I). Furthermore, many of the unidentified birds encountered during the spring were warblers and hummingbirds, many

TABLE 13 Average Density and Daily Energy Expenditure (DEE) of Winter Visitors Recorded During 36 Winter Censuses (Dec.–Feb.) of the Desert Wash Habitat, 1979 and 1980

Species	% Occurrence	Density		DEE	
		birds/40 ha	% of total	kJ/ha	% of total
Phainopepla	77.8	32.05	55.18	57.19	58.70
White-crowned Sparrow	22.2	13.36	23.00	26.98	27.69
Bewick's Wren	36.1	4.15	7.15	3.90	4.01
Ruby-crowned Kinglet	13.9	3.40	5.85	2.66	2.73
Anna's Hummingbird	13.9	2.59	4.46	2.29	2.35
Lesser Goldfinch	2.8	1.40	2.41	1.32	1.35
Blue-gray Gnatcatcher	5.6	0.41	0.71	0.28	0.29
Total of 3 others[a]	–	0.72	1.24	2.82	2.90
TOTAL		58.08	100.00	97.44	100.02

[a]Listed in decreasing density: Rufous-sided Towhee > Rock Wren > Long-eared Owl.

TABLE 14 Average Density and Daily Energy Expenditure (DEE) of Summer Visitors Recorded During 72 Spring and Summer Censuses (Mar.–Aug.) of the Desert Wash Habitat, 1979 and 1980

Species	% Occurrence	Density		DEE	
		birds/40 ha	% of total	kJ/ha	% of total
Scott's Oriole	30.6	3.73	30.37	8.99	31.10
Cliff Swallow	2.8	3.44	28.01	8.83	30.56
Ash-throated Flycatcher	13.9	2.00	16.29	4.04	13.98
Violet-green Swallow	8.3	1.60	13.03	3.29	11.37
Western Kingbird	6.9	0.68	5.54	1.61	5.57
Rough-winged Swallow	5.6	0.63	5.13	1.29	4.48
Lesser Nighthawk	1.4	0.11	0.90	0.31	1.07
White-winged Dove	1.4	0.09	0.73	0.54	1.87
TOTAL		12.28	100.00	28.90	100.00

of the latter migrating *Selasphorus*. Including these unidentified species increases the density of migrants to 86.4 birds/40 ha, or 13.9 percent of all birds detected in spring. Migrants tend to be small compared with resident birds, and they account for only 2.2 percent of the annual energy flow (fig. 27). They may have a greater impact on the bird community than their energy demand indicates, as most migrants are insectivorous and/or nectarivorous. Thus, they may compete with resident species such as Costa's

PLATE 1
Great Horned Owl
Bubo virginianus

PLATE 2 Scott's Oriole *Icterus parisorum*

PLATE 3 Say's Phoebe *Sayornis saya*

PLATE 4 House Finch *Carpodacus mexicanus*

PLATE 5
Black-headed Grosbeak
Pheucticus melanocephalus

PLATE 6 Loggerhead Shrike *Lanius ludovicianus*

PLATE 7 Phainopepla *Phainopepla nitens*

PLATE 8 White-winged Dove *Zenaida asiatica*

TABLE 15 Average Density and Daily Energy Expenditure (DEE) of Spring Migrants Recorded During 48 Spring Censuses (Mar.–May) of the Desert Wash Habitat, 1979 and 1980

Species	% Occurrence	Density		DEE	
		birds/40 ha	% of total	kJ/ha	% of total
Orange-crowned Warbler	35.4	8.77	33.82	8.25	25.75
Yellow Warbler	6.3	3.86	14.89	3.76	11.74
Wilson's Warbler	14.6	3.84	14.81	3.12	9.74
Black-headed Grosbeak	12.5	3.10	11.96	8.34	26.03
Warbling Vireo	8.3	1.70	6.56	1.93	6.02
Hooded Oriole	14.6	0.88	3.39	1.61	5.03
MacGillivray's Warbler	6.3	0.70	2.70	0.75	2.35
Townsend's Warbler	6.3	0.67	2.58	0.61	1.89
Northern Oriole	6.3	0.43	1.66	0.96	3.01
Total of 7 others[a]	–	1.98	7.64	2.70	8.43
TOTAL		25.93	100.01	32.03	99.99

[a]Listed in decreasing density: Vaux's Swift > Yellow-rumped Warbler > Nashville Warbler > Western Flycatcher = Black-throated Gray Warbler > Black Phoebe > Green-tailed Towhee.

Hummingbirds, Verdins, and Black-tailed Gnatcatchers (see chap. 3). I have seen both Costa's Hummingbirds and Verdins attack warblers, suggesting that a thorough study of how migrants affect resident breeding-success might be revealing.

SCRUBLAND

Cacti, low shrubs, stunted trees, and rocky terrain characterize the alluvial plain's scrubland habitat. The prominent plants include shrubs such as creosote bush, burrobush, brittlebush (*Encelia farinosa*), cheesebush (*Hymenoclea salsola*), and several species of prickly-pear and cholla belonging to the genus *Opuntia*. Scattered palo verde, smoke trees, and ocotillo are less numerous here than elsewhere but are important scrubland species nonetheless. Scrubland plant cover increases from the valley floor toward the base of the mountains. In some areas, the landscape changes from starkly barren to relatively lush (as in fig. 28) over a short distance.

Scrubland Bird Censuses

Four scrubland habitat strip transects have been established within 3 km of the mountain's base at elevations between 200 and 300 m (Sections 9, 16, 17; R. 6 S., R. 6 E., Palm Desert Quadrangle). They range in length from 1.0 to 1.48 km (mean 1.21 km), and two pass within 200 m of a

FIGURE 28 Near the mountains' base, the scrubland habitat is fairly lush, with barrel cacti (*Ferocactus acanthodes*), jumping cholla, and tall, wandlike ocotillo interspersed among creosote bushes and burrobushes. Little San Bernardino Mountains in the distant background.

permanent water source at Boyd Research Center. Because Gambel's Quail, House Finch, Black-throated Sparrow, and Mourning Dove are attracted to the water, their transect density exceeds that of areas of waterless scrubland.

The large number of scrubland censuses (105) permits bird density to be plotted by both month (fig. 29) and season (fig. 30). These data reveal that density follows an irregular annual cycle, increasing from a December low of 26 birds/40 ha to a March high of 169 birds/40 ha. A second peak in July resulted from the appearance of hatchling and juvenile Gambel's Quail.

Residents

Eighteen species reside year-round in the scrubland. Of these, only the nine most abundant actually nest on the scrubland. The others all nest nearby and use the scrubland for foraging.

Essentially the same resident species occur in the scrubland and desert wash habitats (cf. tables 11 and 16). In both habitats, the House Finch and Gambel's Quail are the most abundant species and account for over 50 percent of all resident birds. Comparing densities in tables 11 and

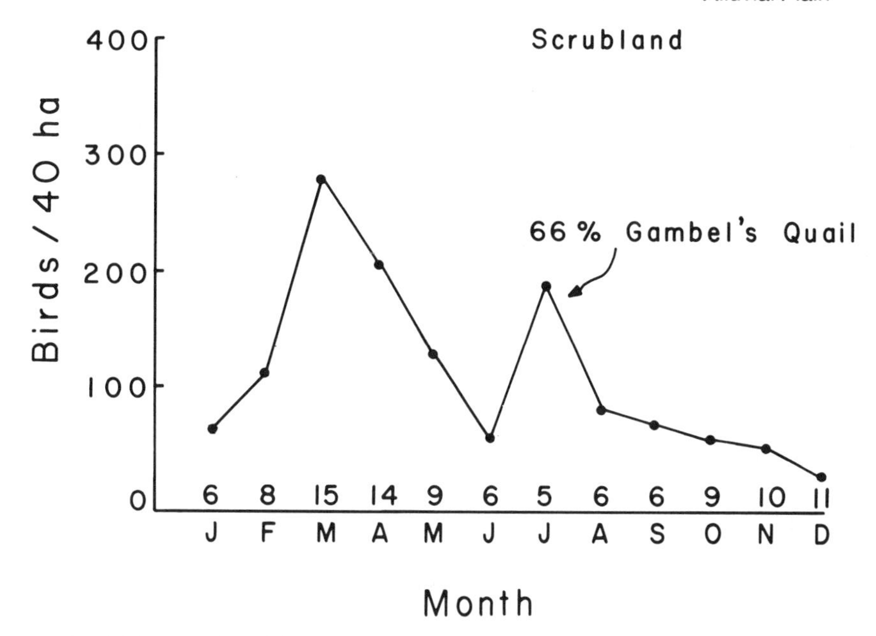

FIGURE 29 Scrubland bird density peaks twice during the year. The July peak results from the appearance of newly hatched Gambel's Quail. Dots are means for the indicated number of censuses.

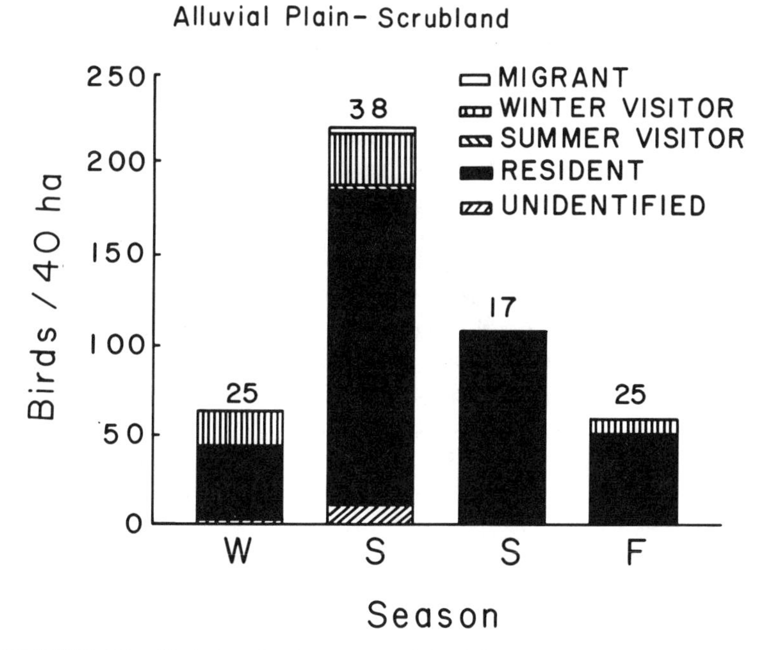

FIGURE 30 Bird density in the scrubland habitat follows the same seasonal pattern as density in the desert wash habitat, but with only one-sixth as many birds (cf. fig. 26). Resident bird species dominate the scrubland habitat.

TABLE 16 Average Yearly Density and Daily Energy Expenditure (DEE) of Permanent Residents Recorded During 105 Censuses of the Scrubland Habitat, 1979 and 1980

Species	% Occurrence	Density		DEE	
		birds/40 ha	% of total	kJ/ha	% of total
House Finch	53.3	35.88	34.94	56.84	16.33
Gambel's Quail	45.7	35.15	34.23	226.11	64.95
Black-throated Sparrow	39.1	8.32	8.10	9.75	2.80
Cactus Wren	44.8	5.48	5.34	13.65	3.92
Costa's Hummingbird	42.9	4.11	4.00	3.03	0.87
Verdin	42.9	4.08	3.97	3.10	0.89
Mourning Dove	20.0	2.87	2.80	14.65	4.21
Black-tailed Gnatcatcher	23.8	2.61	2.54	1.72	0.49
Loggerhead Shrike	25.7	2.51	2.44	6.85	1.97
Say's Phoebe	12.4	0.84	0.82	1.37	0.39
Total of 8 others[a]	–	0.83	0.81	11.07	3.18
TOTAL		102.68	99.99	348.14	100.00

[a]Listed in decreasing density: Red-tailed Hawk > Mockingbird = Common Raven > Cooper's Hawk = Roadrunner > Prairie Falcon > Burrowing Owl > American Kestrel.

16 reveals that, of all the resident species, only the Black-throated Sparrow prefers the scrubland over the desert wash.

Some resident birds exhibit marked seasonal changes in density (table 17), which can be explained according to various factors. In spring, House Finches are attracted to the scrubland by the Boyd Research Center buildings, where they nest colonially. The summer quail peak results from the appearance of large numbers of young and from quail moving across the scrubland to drink at the Boyd Research Center. The summer Cactus Wren increase represents young birds produced on the transect. Declining wren numbers in fall and winter presumably reflect emigration and/or death of young birds. The spring and summer increase in shrikes represents young birds produced on the transect.

Daily energy expenditure by scrubland birds varies 3.7-fold between spring and fall (fig. 31). Resident birds account for over 90 percent of the total energy flow in summer and fall but only 79 and 87 percent in winter and spring, respectively. This reflects the influx of winter visitors and migrants. On an annual basis, however, resident birds account for 93 percent of the community's total energy expenditure.

Winter Visitors

Winter visitors were second in importance to resident birds in terms of both density and energy flow (figs. 30 and 31). They accounted for 32 percent of the birds and 20 percent of the energy expenditure during

TABLE 17 Seasonal Change in Resident Bird Density in the Scrubland Habitat

	Density (birds/40 ha)			
	Winter	Spring	Summer	Fall
House Finch	7.7	90.0	3.6	3.8
Gambel's Quail	13.2	46.4	70.1	16.2
Cactus Wren	1.1	6.2	11.7	4.6
Loggerhead Shrike	1.2	4.2	3.3	0.8

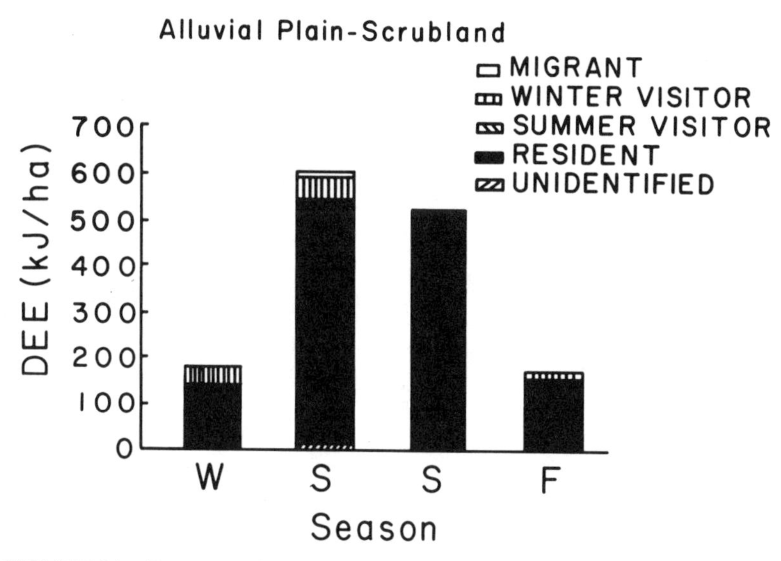

FIGURE 31 Permanently resident species dominate avian daily energy expenditure (DEE) in the scrubland habitat. The second most important seasonal group (winter visitors) accounts for only 6.7 percent of the total annual energy flow.

winter. As in the adjacent desert washes, the most abundant scrubland winter visitors were the White-crowned Sparrow and the Phainopepla (table 18). The density of winter visitors in the scrubland was much lower than in the desert wash (20.26 versus 58.08 birds/40 ha), reflecting the lower density of scrubland vegetation.

Relatively few birds find the desert's scrubland attractive in winter, and only the Rock Wren prefers this habitat at that time. Nineteen species, including residents and migrants, were recorded during winter scrubland censuses, and most were represented by just a few individuals. Consequently, winter bird density was a mere 63 birds/40 ha. Four granivorous species account for 57 percent of all birds encountered: Gambel's Quail, White-crowned Sparrow, House Finch, and Black-throated Sparrow. Their recorded abundance may be an overestimate for the scrubland as a whole, however, because some of the census sites were located where these species congregated for water.

TABLE 18 Average Density and Daily Energy Expenditure (DEE) of Winter Visitors Recorded During 25 Winter Censuses (Dec.–Feb.) of the Scrubland Habitat, 1979 and 1980

Species	% Occurrence	Density		DEE	
		birds/40 ha	% of total	kJ/ha	% of total
White-crowned Sparrow	24.0	9.65	47.63	19.49	53.76
Phainopepla	40.0	5.30	26.16	9.46	26.09
Violet-green Swallow	4.0	1.65	8.14	3.39	9.35
Rock Wren	24.0	1.42	7.01	1.95	5.37
Anna's Hummingbird	12.0	0.96	4.74	0.85	2.34
Bewick's Wren	8.0	0.96	4.74	0.90	2.49
Blue-gray Gnatcatcher	4.0	0.32	1.58	0.22	0.60
TOTAL		20.26	100.00	36.26	100.00

TABLE 19 Average Density and Daily Energy Expenditure (DEE) of Spring Migrants Recorded During 38 Spring Censuses (Mar.–May) of the Scrubland Habitat, 1979 and 1980

Species	% Occurrence	Density		DEE	
		birds/40 ha	% of total	kJ/ha	% of total
Chipping Sparrow	7.9	0.71	28.63	0.81	24.02
Western Wood Pewee	5.3	0.28	11.29	0.32	9.47
Total of 9 others[a]	–	1.49	60.08	2.23	66.52
TOTAL		2.48	100.00	3.36	100.01

[a]Listed in decreasing density: Black-headed Grosbeak > Rufous Hummingbird = Wilson's Warbler = Yellow Warbler > Western Tanager > Hooded Oriole = Northern Oriole = Orange-crowned Warbler > Warbling Vireo.

Summer Visitors

In summer the scrubland is hot, dry, and singularly uninviting. Of the summer visitors known to occur on the Deep Canyon Transect, only the Lesser Nighthawk nests in the scrubland. Only four summer visitors were recorded during the summer censuses: Ash-throated Flycatcher, White-winged Dove, Lesser Nighthawk, and Scott's Oriole. Their combined density was a mere 1.36 birds/40 ha.

Migrants

Spring migrants find the scattered ocotillo and palo verde of the scrubland less attractive than the lush desert wash vegetation, and consequently they are much less numerous on the scrubland (cf. tables 15 and 19). Spring migrants and unidentified birds (mostly hummingbirds and warblers) in the scrubland have a combined density of only 11.84 birds/40 ha (versus 86.42/40 ha in the desert wash). This further emphasizes the importance of the washes to migrating birds.

seven

ROCKY SLOPES

These ranges are as nearly barren and desolate as one could well imagine, their scarred flanks reminding one of sun-mummied carcasses.

—A. Brazier Howell

Deep Canyon's alluvial plain ends abruptly at the base of the Santa Rosa Mountains, giving way to steep, rocky slopes that ascend to the lower plateau. The landscape suddenly becomes one of unrelieved rock—weathered and fractured into a loose debris—talus slides alternating with rocky spurs and ridges (fig. 32). Summer air temperatures on the rocky slopes frequently exceed 40°C (104°F), and the nearest water is often miles away. The stunted and widely spaced plants evoke a false sense of desolation and suggest that birds could not possibly find this barren habitat appealing. But, visit the rocky slopes and you will find birds, even in midsummer when the withered shrubs seem lifeless and heat waves shimmer all around.

FLORA

Poor soil, high temperatures, lack of water, and sparse vegetation all imply that few plant species inhabit the rocky slopes. The slopes' vegetation is remarkably rich, however, and includes at least 102 perennial and 115 annual species (Zabriskie 1979). The more common perennials include brittlebush, burrobush, creosote bush, and desert agave (*Agave deserti*). Together they comprise 62 percent of the plant cover (Zabriskie 1979). Other important perennials are ocotillo, desert lavender, and pigmy cedar (*Peucephyllum schottii*).

The agave is one of the rocky slopes' more striking plants, with its compact, basal rosette of spearlike leaves and single, large, stalked inflorescence. Each rosette flowers only once during its life, following a period of years during which it gradually accumulates the energy needed to produce its massive flower stalk. Why the agave should invest so much energy in flowering is something of an enigma, as new plants apparently develop vegetatively from the base of older plants and seedlings rarely survive.

FIGURE 32 The distinctive plants of the rocky slopes habitat include the rosette-leaved desert agave (foreground) and the stately barrel cactus. Rock Wrens reach their highest density on these rocky slopes.

The agave is an important food source for many desert animals. Desert bighorn sheep (*Ovis canadensis*) and several rodent species eat the agave's leaves. In spring, its waxy, yellow flowers attract numerous orioles and warblers. Deep Canyon's original settlers, the Cahuilla Indians, relied heavily on the agave for both food and fiber. They baked its leaves, stalk, and heart in stone pits and either boiled the blossoms or ate them raw (Wilke 1976). The numerous baking pits that punctuate Deep Canyon's hillsides attest to the agave's importance to the Cahuilla.

BIRD CENSUSES

Two transects have been established in the rocky slopes habitat. One is a standard strip transect (i.e., 50 m wide) that ascends Worley Trail a distance of 0.66 km (Section 17; T. 6 S., R. 6 E., Palm Desert Quadrangle). The other is rectangular, measuring 0.54 by 0.30 km (Sections 7, 8, 17, 18; T. 6 S., R. 6 E., Palm Desert Quadrangle). This latter transect cuts across several steep ravines and talus slopes that make walking arduous. The perennial plants are similar on both transects, with a few exceptions. Agaves are numerous on the strip transect but scarce on the rectangular transect. Ocotillo is dense on the rectangular transect but scarce on the strip transect.

Bird density on the rocky slopes is very low and shows less seasonal change than elsewhere at Deep Canyon (see fig. 12). The mean annual bird density, a paltry 105 birds/40 ha, is the lowest of all the habitats, which reflects the rocky slopes' harsh climate and limited vegetation. The minimum mean monthly density (32 birds/40 ha in December) is equivalent to one bird in an area the size of three football fields. Even the maximum mean monthly density (169 birds/40 ha in May) is so low that a bird watcher must patiently survey the area to glimpse any bird activity at all. Resident species dominate this sparse avifauna and account for 90 percent of all the birds (fig. 33).

Residents

In spite of the low bird density, the diversity of resident species is similar to that of the adjacent scrubland habitat (table 20). Most of the resident species are the same here as on the scrubland. They also nest in the same plant species as on the scrubland, although the plants themselves are fewer and farther apart. Some residents nest on the cliffs. Raptors and ravens nest on the remote cliff ledges, whereas White-throated Swifts nest in the crevices and exfoliating rock of the canyon walls.

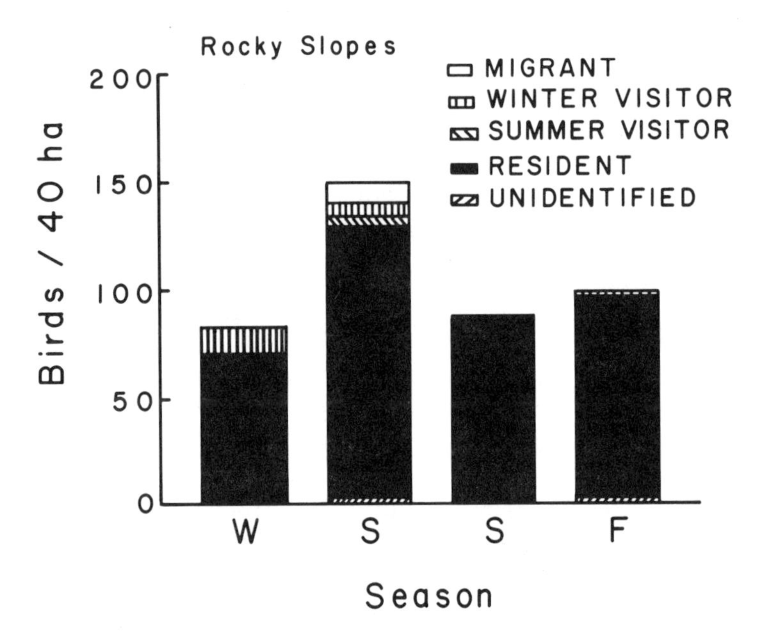

FIGURE 33 Of all Deep Canyon's habitats, the rocky slopes has the fewest birds and shows the smallest seasonal change in bird density. Resident species dominate these desert slopes.

TABLE 20 Average Yearly Density and Daily Energy Expenditure (DEE) of Permanent Residents Recorded During 58 Censuses of the Rocky Slopes Habitat, 1978 through 1980

Species	% Occurrence	Density		DEE	
		birds/40 ha	% of total	kJ/ha	% of total
Black-throated Sparrow	86.2	31.36	32.26	36.74	16.84
House Finch	75.9	20.46	21.05	32.41	14.85
Costa's Hummingbird	67.2	9.50	9.77	7.00	3.21
Rock Wren	72.4	8.44	8.68	11.56	5.30
Mourning Dove	53.5	7.96	8.19	40.63	18.62
Gambel's Quail	25.9	3.65	3.75	23.48	10.76
Loggerhead Shrike	53.5	3.30	3.39	9.01	4.13
Bewick's Wren	43.1	2.87	2.95	2.70	1.24
Cactus Wren	46.6	2.53	2.60	6.30	2.89
White-throated Swift	12.1	1.87	1.92	6.21	2.85
Verdin	36.2	1.58	1.63	1.20	0.55
Say's Phoebe	31.0	1.11	1.14	1.82	0.83
Red-tailed Hawk	19.0	0.89	0.92	18.88	8.65
Common Raven	15.5	0.73	0.75	13.52	6.20
American Kestrel	6.9	0.59	0.61	3.00	1.37
Roadrunner	10.3	0.26	0.27	2.41	1.10
Total of 3 others[a]	–	0.12	0.12	1.34	0.61
TOTAL		97.22	100.00	218.21	100.00

[a]Each having a density of 0.04 birds/40 ha: Cooper's Hawk, Great Horned Owl, Ladder-backed Woodpecker.

Resident species of the rocky slopes tend to be insectivorous and/or carnivorous. Granivorous birds are relatively scarce and account for only 33 percent of the total bird density and 44 percent of daily energy expenditure. What factor(s) might contribute to this dearth of seed eaters? Low seed availability seems unlikely, since two common perennial plants, brittlebush and burrobush, produce abundant seed crops. Furthermore, granivorous rodents attain their highest density (5,900–7,492 rodents/40 ha) in this habitat (Ryan 1968), suggesting that seeds are indeed abundant. Interestingly, the rodent density is 100 times higher than the granivorous bird density. Since rodents are better able to locate seeds than are birds and may remove most of the surface seeds, perhaps competition with rodents suppresses bird numbers (but see chapter 3).

The rocky slopes' most abundant bird is the Black-throated Sparrow, formerly known as the Desert Sparrow. It nests in low shrubs, such as brittlebush and burrobush and during winter eats mainly seeds. In spring and summer its diet changes to insects, which supply enough water to make the sparrows independent of drinking water, even during summer's severest heatwaves (Smyth and Bartholomew 1966).

Extreme environments are typically dominated by a few relatively abundant species. The rocky slopes habitat conforms to this pattern, with five species accounting for 80 percent of the total resident bird density (table 20). These numerically dominant species account for proportionately less (59 percent) of the estimated energy expenditure by resident birds than their density would suggest. Thus, this habitat exhibits greater functional evenness than its structure suggests.

Several resident species exhibit marked seasonal changes in density (table 21). Some of these changes have obvious explanations; others do not. In the spring, Costa's Hummingbirds, for example, are attracted to the rocky slopes by the flowering ocotillo (fig. 34). The spring increase in Black-throated Sparrow density reflects the arrival of breeding pairs from other regions of the Deep Canyon Transect. The reasons for the seasonal density changes in the other species are unknown.

Resident birds dominate energy flow in this habitat, accounting for 93 percent of the annual energy flow of 85.5 MJ/ha. As with bird density, daily energy expenditure peaks in the spring (fig. 35). Unlike density, however, it shows a smaller decline in summer and fall owing to an influx of large birds such as the Mourning Dove (table 21).

Winter Visitors

Of all Deep Canyon's habitats, only the lower plateau and coniferous forest had fewer species of winter visitors than the rocky slopes (see table 5). A mere three species were detected during seventeen winter censuses (table 22). Two of the three are insectivorous, which indicates that, as with resident species, seed-eating winter visitors find these slopes unappealing. For example, the granivorous White-crowned Sparrow, the second most abundant winter visitor in the desert wash habitat, was never encountered on the rocky slopes.

Winter visitors rank a distant second to residents in terms of both density and energy flow (figs. 33 and 35), accounting for only 2.6 percent of the annual energy flow and 14 percent of the winter bird density.

TABLE 21 Seasonal Changes in Resident Bird Density in the Rocky Slopes Habitat

	Density (birds/40 ha)			
	Winter	Spring	Summer	Fall
Black-throated Sparrow	27.56	40.52	22.72	27.57
House Finch	15.81	18.76	14.89	32.57
Costa's Hummingbird	6.27	19.37	4.06	1.89
Mourning Dove	2.62	12.31	20.74	0.38
Loggerhead Shrike	1.61	6.96	1.85	0.77

FIGURE 34 In spring, blossoms of the ocotillo attract hummingbirds and warblers to the rocky slopes. Black-throated Sparrows nest in the scattered burrobushes and brittlebushes that dot the hillside.

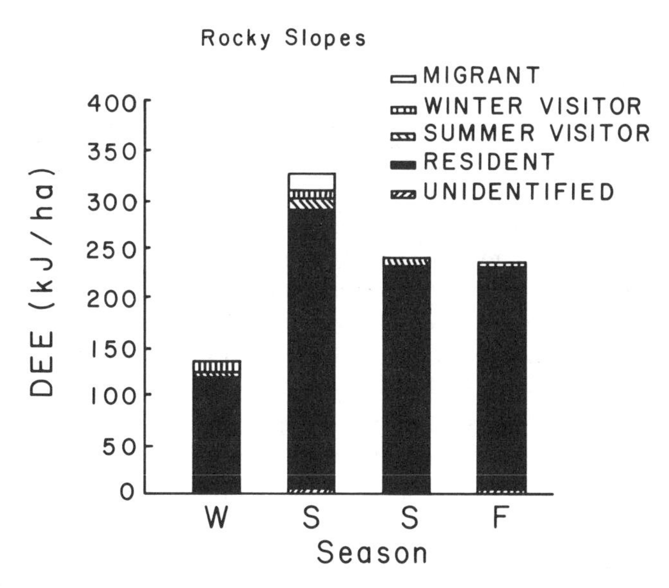

FIGURE 35 The rocky slopes habitat is dominated by resident species, which account for 93 percent of the total daily energy expenditure (DEE).

TABLE 22 Average Density and Daily Energy Expenditure (DEE) of Winter Visitors Recorded During 17 Winter Censuses (Dec.–Feb.) of the Rocky Slopes Habitat, 1978 through 1980

Species	% Occurrence	Density		DEE	
		birds/40 ha	% of total	kJ/ha	% of total
Yellow-rumped Warbler	35.3	8.67	53.85	9.85	60.47
Lesser Goldfinch	29.4	3.92	24.35	3.69	22.65
Ruby-crowned Kinglet	41.2	3.51	21.80	2.75	16.88
TOTAL		16.10	100.00	16.29	100.00

TABLE 23 Average Density and Daily Energy Expenditure (DEE) of Summer Visitors Recorded During 28 Spring and Summer Censuses (Mar.–Aug.) of the Rocky Slopes Habitat, 1978 through 1980

Species	% Occurrence	Density		DEE	
		birds/40 ha	% of total	kJ/ha	% of total
Scott's Oriole	46.4	1.77	45.27	4.26	43.29
Ash-throated Flycatcher	32.1	1.34	34.27	2.71	27.54
Western Kingbird	17.9	0.53	13.55	1.25	12.70
White-winged Dove	7.1	0.27	6.91	1.62	16.46
TOTAL		3.91	100.00	9.84	99.99

Summer Visitors

Four species of summer visitors were found on Deep Canyon's rocky slopes (table 23). Their numbers were quite low, however, and collectively they accounted for only 3.3 percent of the birds and 3.9 percent of the daily energy expenditure during their spring peak (figs. 33 and 35). In summer these values declined to 1.4 percent and 1.2 percent respectively, indicating that some of the spring birds were actually migrants. This was obviously true of the Scott's Oriole. In spring, Scott's Orioles were attracted to the rocky slopes by the ocotillo blossoms, and their density averaged 2.36 orioles/40 ha at that time. In summer, after the ocotillo had finished flowering, their density dropped to only 0.31 birds/40 ha. The rocky slopes are marginal nesting habitat for this species, and no nests were found there.

Migrants

A profusion of flowers makes the rocky slopes surprisingly colorful in spring, and a morning spent near a flowering stand of agave or ocotillo will yield rewarding views of many bird species. In March and April migrant

TABLE 24 Average Density and Daily Energy Expenditure (DEE) of Spring Migrants Recorded During 20 Spring Censuses (Mar.–May) of the Rocky Slopes Habitat, 1978 through 1980

Species	% Occurrence	Density		DEE	
		birds/40 ha	% of total	kJ/ha	% of total
Barn Swallow	10	1.73	18.04	4.16	24.30
Nashville Warbler	35	1.50	15.64	1.32	7.71
Western Tanager	5	1.12	11.68	2.31	13.49
Black-headed Grosbeak	30	0.88	9.18	2.37	13.84
Orange-crowned Warbler	25	0.75	7.82	0.71	4.15
Black-throated Gray Warbler	10	0.50	5.21	0.45	2.63
Western Wood Pewee	15	0.38	3.96	0.43	2.51
Cassin's Kingbird	15	0.38	3.96	1.04	6.07
Total of 12 others[a]	–	2.35	24.50	4.33	25.29
TOTAL		9.59	99.99	17.12	99.99

[a]In order by decreasing density: Wilson's Warbler = Lazuli Bunting > Western Flycatcher = Rough-winged Swallow = Vaux's Swift > Cliff Swallow > Merlin = Warbling Vireo = Northern Oriole = Hammond's Flycatcher = Townsend's Warbler = Pine Siskin.

Nashville and Orange-crowned Warblers drift across the rocky slopes and linger briefly at the crimson ocotillo blossoms. Occasional waves of Western Tanagers appear, adding a splash of orange and black to the desert agave's yellow flowers. The rocky slopes host at least twenty species of spring migrants (table 24), and in some years the number is undoubtedly higher.

Migrants account for 7.6 percent of the individuals (fig. 33) and 5.2 percent of the total daily energy expenditure in spring (fig. 35). Many of the unidentified birds encountered in spring were warblers and *Selasphorus* hummingbirds. If their density (2.69 birds/40 ha) is added to that of the migrants, the total migrant bird density becomes 12.28 birds/40 ha, or 8.2 percent of the total springtime bird density (fig. 33).

eight

LOWER PLATEAU

At an elevation of 800 m, Deep Canyon's steep rocky slopes flatten out onto a gently rising plateau that extends southward 12 km. The plateau is the upper limit of the creosote bush scrub community, and it is here that the transition from low desert to high mountain habitat occurs. Succulent and drought-deciduous desert vegetation on the lower half of the plateau merges through an ecotone of yucca, juniper (*Juniperus californica*), and scrub oak (*Quercus turbinella*) into a mature piñon woodland at the plateau's upper end.

FIGURE 36 Mojave yucca and desert agave flower stalks rise above the lower plateau's cacti. House Finches, which nest colonially in the abundant jumping cholla cactus, make this habitat second only to the coniferous forest in total nest density. This is the preferred habitat of the Ladder-backed Woodpecker (see frontispiece).

Near its junction with the rocky slopes, the lower plateau is a region of spiniferous succulents and stunted shrubs, of rocky outcrops that alternate with shallow washes and sandy flats. The generally rocky terrain is dominated by agave, ocotillo, Mojave yucca (*Yucca schidigera*), galleta grass (*Hilaria rigida*), burrobush, brittlebush, and several species of cacti (fig. 36). It resembles the southern edge of the Mojave Desert—located across the Coachella Valley—with which it shares many plant and animal species. Deep Canyon Gorge cuts through the plateau, and its sheer cliffs are the nesting sites of Golden Eagles, White-throated Swifts, Common Ravens, and Prairie Falcons.

Ecologically, the lower plateau is an extension of the scrubland and rocky slopes habitats. It closely resembles them in dominance of desert-adapted species and low bird density. The low vegetation is floristically more diverse than that of the lower elevations but is less variable in size and structure. This lack of vegetative variability reduces the choice of nesting sites, and consequently few bird species nest here. Still, a few species find these nesting conditions ideal, and two, the Black-throated Sparrow and House Finch, nest in abundance. Accordingly, the lower plateau ranks second only to the coniferous forest in nest density (see table 5).

BIRD CENSUSES

Two strip transects, 0.71 and 0.36-km long, have been established on the lower plateau (Section 19; T. 6 S., R. 6 E., Palm Desert Quadrangle). Both pass through habitat like that pictured in figure 36.

In winter, bird density on the lower plateau was extremely low (fig. 37), averaging a scant 17.76 birds/40 ha, the lowest winter density of any habitat (see fig. 12). No species of winter visitors were found, and on several winter censuses no birds were found. In spring, bird density increased over 10-fold to 235.76 birds/40 ha. This increase was due almost entirely to an influx of House Finches and Black-throated Sparrows. In large part, these two species tell the density-diversity story of the lower plateau, and listing their density by season (table 25) clearly illustrates their role in the yearly cycle depicted in figure 37.

In spring, House Finches inundate the lower plateau. They nest in loose colonies in the abundant jumping cholla (*Opuntia bigelovii*) and become almost as abundant there as in the desert washes (fig. 38). In summer, they leave the lower plateau, while Black-throated Sparrows remain behind to re-nest (table 25). This difference in departure time is curious because Black-throated Sparrows are about half the size of House Finches and thus should be less tolerant of heat, other things being equal. What enables Black-throated Sparrows to tolerate this truly scorching summer home when the larger House Finches must leave? The answer seems to be diet and behavior.

During summer Black-throated Sparrows eat mostly insects (rich in water) and are seldom seen at water holes. In contrast, the more graniv-

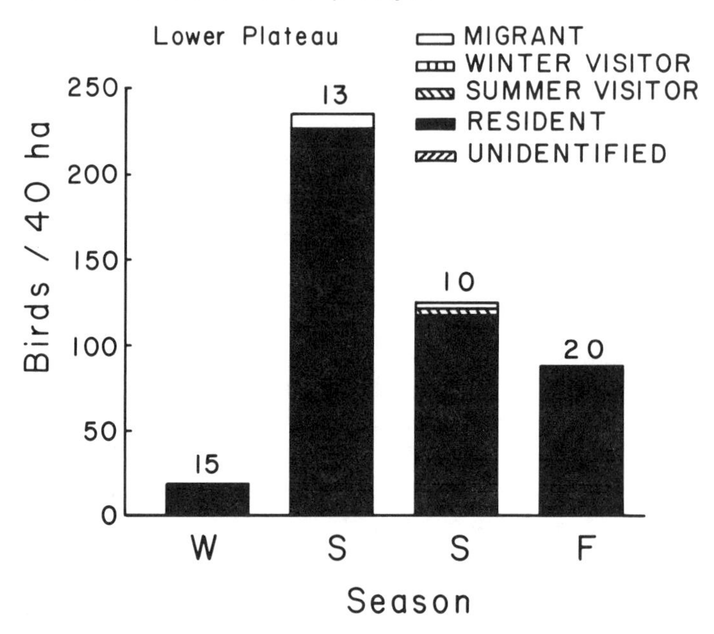

FIGURE 37 During winter, bird density on the lower plateau is negligible. House Finches account for 65 percent of the birds found in the spring. See fig. 16 for explanation of symbols.

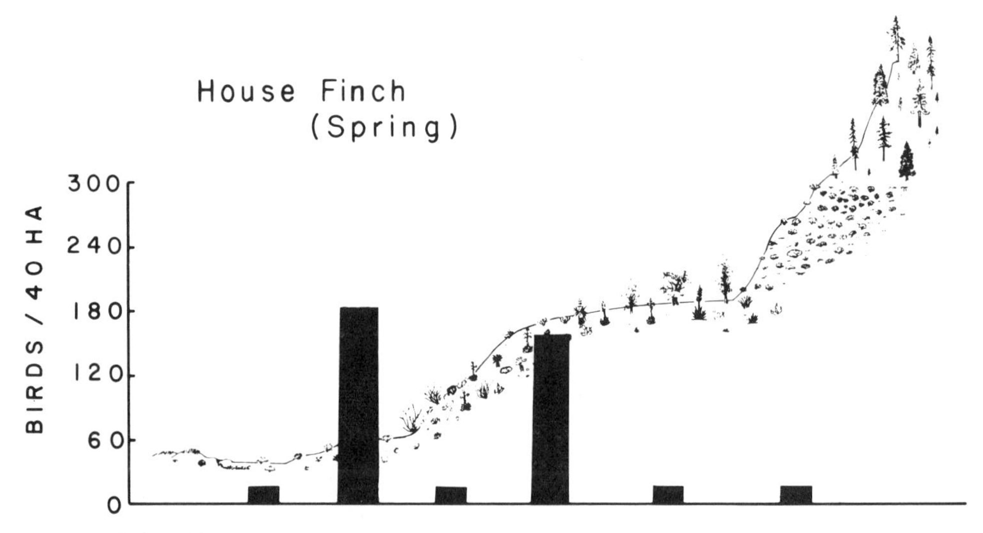

FIGURE 38 In spring, House Finches concentrate on the lower plateau and alluvial plain. The vertical bars represent the mean density during the indicated time period. They are centered beneath the habitat to which they refer. (See fig. 3 for details of the Deep Canyon profile.)

orous House Finch requires drinking water during the summer, but none is available on the lower plateau. As summer approaches and the days become hotter, House Finches lose even more water by evaporation.

TABLE 25 Seasonal Change in Bird Density on the Lower Plateau

	Density (birds/40 ha)			
	Winter	Spring	Summer	Fall
House Finch	1.48	154.84	13.40	10.54
Black-throated Sparrow	2.22	46.42	52.34	43.40

FIGURE 39 Typical House Finch nest in jumping cholla. The cactus provides protection from some predators (e.g., biologists) but may leave the young exposed to the desert sun.

Their demand for water may exceed the available supply, thereby forcing their departure from the dry plateau.

Although this explanation is attractive, differences in nest site selection may be even more important than diet. Black-throated Sparrows typically place their nests in dense burrobushes or brittlebushes. Such sites are more protected from the intense sun than are the open nest sites of the House Finch (fig. 39). But even in their more protected sites, Black-throated Sparrows are subject to heat stress. When air temperatures are high, adult sparrows perch quietly on the nest's edge and pant vigorously while shading their young from the sunlight that filters through the foliage. Panting requires water, and were it not for the water obtained from their diet, Black-throated Sparrows would be forced to leave the lower plateau along with the House Finches. Hence, the Black-throated Sparrow's ability to breed on the lower plateau in summer rests upon a combination of behavior and diet.

Residents

Five of Deep Canyon's resident species reach their highest density on the lower plateau: the Black-throated Sparrow, White-throated Swift, American Kestrel, Ladder-backed Woodpecker, and Golden Eagle. All five species are attracted to the lower plateau by an abundance of nest sites. The low, dense vegetation favored by the Black-throated Sparrow is more extensive here than elsewhere on the Deep Canyon Transect and attracts numerous nesting sparrows. The Ladder-backed Woodpecker (see frontispiece) nests in cavities that it excavates in the dead flower stalks of the desert agave. The stalks seemingly persist for many years, as old nest cavities are much more common than woodpeckers. The eagle, swift, and kestrel nest on the canyon walls and forage across the plateau.

Fifteen species reside permanently on the lower plateau (table 26). The Black-throated Sparrow and House Finch alone represent 72 percent of the individual residents. Because of their small size, however, they only account for 39 percent of the average daily energy expenditure by resident birds. The Common Raven, although only one-tenth as numerous as the House Finch, is much larger and consequently has a DEE exceeding that of the House Finch

TABLE 26 Average Yearly Density and Daily Energy Expenditure (DEE) of Permanent Residents Recorded During 58 Censuses of the Lower Plateau Habitat, 1979 and 1980

Species	% Occurrence	Density		DEE	
		birds/40 ha	% of total	kJ/ha	% of total
House Finch	36.2	41.03	38.88	65.00	23.91
Black-throated Sparrow	58.6	34.97	33.14	40.97	15.07
White-throated Swift	13.8	6.54	6.20	21.73	7.99
Rock Wren	27.6	5.93	5.62	8.12	2.99
Common Raven	12.1	3.83	3.63	70.97	26.10
Cactus Wren	10.3	3.26	3.09	8.12	2.99
Costa's Hummingbird	10.3	1.93	1.83	1.42	0.52
Loggerhead Shrike	12.1	1.72	1.63	4.69	1.73
American Kestrel	6.9	1.34	1.27	6.80	2.50
Mourning Dove	8.6	1.15	1.09	5.87	2.16
Ladder-backed Woodpecker	8.6	1.15	1.09	2.68	0.98
Red-tailed Hawk	5.2	0.96	0.91	20.36	7.49
Gambel's Quail	3.5	0.77	0.73	4.95	1.82
Say's Phoebe	5.2	0.76	0.72	1.24	0.46
Golden Eagle	1.7	0.19	0.18	8.95	3.29
TOTAL		105.53	100.01	271.87	100.00

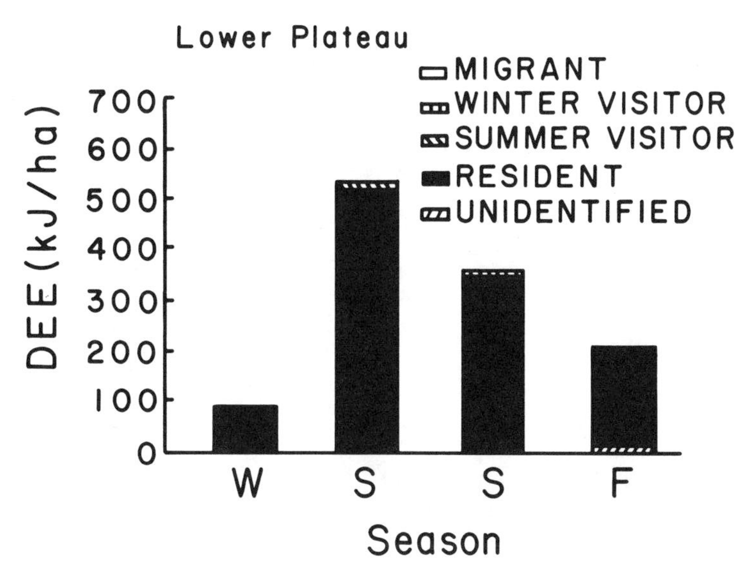

FIGURE 40 The seasonal change in average daily energy expenditure (DEE) by lower plateau birds closely parallels the seasonal change in bird density. Residents account for 99 percent of the annual energy flow.

The seasonal change in average DEE parallels the change in bird density (fig. 40). Energy flow shows a proportionately smaller spring bulge than density, however, increasing to only 5.7 times the winter value, versus 13.3 times for the density value. Thus, the lower plateau's bird community is functionally more stable than the density change suggests.

As is the case in the rocky slopes habitat, many of the lower plateau's resident birds are either insectivorous or carnivorous (table 26). This suggests that animal prey is more available than are seeds.

Summer Visitors

Scott's Orioles were the only summer visitors found on the lower plateau. They built their fibrous, hanging nests in the Mojave yuccas and had a mean density of 0.98 orioles/40 ha. This insectivore's diet supplies all the water it needs to survive the desert summer.

Migrants

Migrant species rank a distant second to residents in both density and energy flow, accounting for a mere 2.6 percent of all springtime birds. Only four species of migrants were recorded during spring censuses of the lower plateau (table 27), much lower than the 20 species found on the rocky slopes transects. This difference between the two habitats may result from the much higher ocotillo density on the rocky slopes. In spring, the ocotillo's crimson blossoms attract a variety of warblers and

TABLE 27 Average Density and Daily Energy Expenditure (DEE) of Spring Migrants Recorded During 13 Spring Censuses (Mar.–May) of the Lower Plateau Habitat, 1979 and 1980

Species	% Occurrence	Density		DEE	
		birds/40 ha	% of total	kJ/ha	% of total
Lesser Goldfinch	7.7	1.71	33.14	1.61	23.54
Lazuli Bunting	7.7	1.71	33.14	2.15	31.43
Barn Swallow	7.7	0.87	16.86	2.09	30.56
Yellow-rumped Warbler	7.7	0.87	16.86	0.99	14.47
TOTAL		5.16	100.00	6.84	100.00

other insectivores. The bright flowers, conspicuous against the darkly varnished rock, are larger and more numerous on the rocky slopes than on the lower plateau.

nine

PIÑON-JUNIPER WOODLAND

The piñon-juniper community is unique to the arid mountains of the western United States, where it ranges from Utah and New Mexico through California to Baja California. At Deep Canyon, piñon-juniper woodland covers the slopes of the lesser peaks, such as Sugarloaf, Sheep, and Martinez mountains, and forms a discontinuous cover across the upper plateau. It lies at the junction between the cool, moist, coastal climate and the hot, dry, interior climate and marks the transition between the low-lying Colorado Desert and the high mountain forests. Elements of both climatic regions occur here and contribute to the piñon-juniper's diversity. The vegetation is more variable in height and size on the upper plateau than at the lower elevations. Piñon trees (*Pinus monophylla*) rise above a mixed understory of evergreen shrubs (fig. 41). Nevertheless, the sandy soil and presence of succulents, such as golden cholla (*Opuntia echinocarpa*) and pancake cactus (*O. chlorotia*), are constant reminders that this is arid habitat. The dense vegetation covers 41 percent of the ground surface at 1,220 m on the plateau, and 92 percent of the plants are evergreens (Zabriskie 1979). The broken understory is dominated by manzanita (*Arctostaphylos glauca*), sugar bush (*Rhus ovata*), scrub oak, juniper, desert apricot (*Prunus fremontii*), and nolina (*Nolina parryi*). Deep Canyon's grasses reach their highest diversity and greatest abundance in the piñon-juniper woodland. It is thus no surprise that half of all birds encountered here are granivorous. Seventeen of the twenty-five grass species that occur in the piñon-juniper are perennial (Zabriskie 1979), with species of the genera *Aristida* and *Bouteloua* being dominant.

BIRD CENSUSES

Five strip transects ranging in length from 0.75 to 1.0 km (mean 0.83 km) have been established on Pinyon Flat (Section 34; T. 6 S., R. 5 E., Palm Desert Quadrangle). One hundred censuses were conducted at fairly even intervals throughout the year. Thus, this habitat was censused an average of once every three days.

FIGURE 41 Piñon trees, scrub oak, and juniper are dominant plants of the piñon-juniper woodland. Several species of cacti and the *Nolina* (left of photo) increase plant diversity.

Bird density in the piñon-juniper exhibits a unique seasonal pattern (fig. 42), with the greatest number of birds occurring during fall. The density drops slightly during winter, plummets in spring, and decreases even farther during summer. This odd pattern results from the seasonal change in numbers of residents and winter visitors.

Although the spring decline in winter visitors results from the departure of several species, the change in resident bird density is primarily because of just three species: the Common Raven, Mourning Dove, and Rufous-sided Towhee (table 28). If the totals from table 28 for these three species are subtracted from the resident columns in figure 42, the amplitude of the seasonal density change is greatly reduced, and the average yearly density of residents becomes 110.60 birds/40 ha (range: 92.65–136.18). Most resident bird populations of the piñon-juniper are thus more stable than figure 42 suggests.

What accounts for the seasonal population changes in the three resident species? Density changes for Mourning Dove and Rufous-sided Towhee may result from local population shifts. The fall-winter decline in towhee numbers in the chaparral coincides with an increase in their numbers in the piñon-juniper woodland. Similarly, Mourning Dove numbers increase in the piñon-juniper as they decrease at the lower elevations. No firm evidence exists that such local population shifts occur, however, and the winter influx of towhees and doves into the piñon-juniper may reflect the arrival of northern migrants.

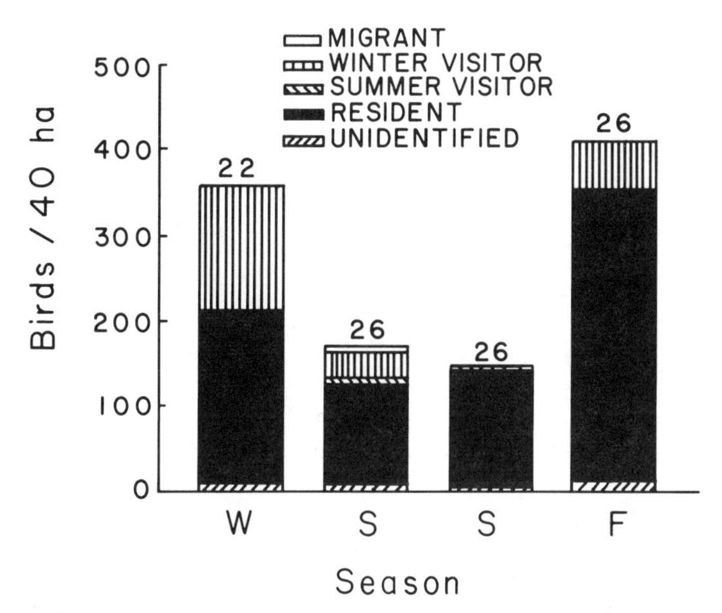

FIGURE 42 Bird density in the piñon-juniper woodland was higher in winter and fall than in spring and summer—a radical departure from the usual pattern. Winter visitors were important components of the avifauna, and several species seemed to migrate down-mountain from the coniferous forest to winter among the piñon trees. See fig. 16 for explanation of symbols.

Common Ravens do not show a similar local habitat shift. They are present in the piñon-juniper habitat in low numbers most of the year. But from late September to late October in 1979 through 1981, large flocks (some containing over 125 individuals) suddenly appeared. Their arrival coincided with the availability of piñon pine cones. Mayhew (personal communication) observed the ravens feeding on the pine cones, suggesting that this resource had attracted the enormous flocks.

Average daily energy expenditure (DEE), like bird density, exhibits a prominent fall peak (fig. 43). This peak also reflects the high fall density of the Common Raven. Ravens (average mass 864 grams) are much larger than other birds of the piñon-juniper, and consequently they require more energy per bird. Ravens account for 78 percent of the fall DEE and 51 percent of the annual energy used by birds in the piñon-juniper habitat.

Residents

The piñon-juniper community contains twenty-six resident bird species, an unsurpassed number among Deep Canyon's habitats. This great species diversity presumably results from this habitat's vegetational complexity, which provides a wide range of foraging and nesting opportunities.

TABLE 28 Seasonal Change in Density of Resident Birds in the Piñon-juniper Habitat

	Density (birds/40 ha)			
	Winter	Spring	Summer	Fall
Mourning Dove	66.35	24.37	45.73	49.90
Common Raven	1.75	2.17	0.00	145.38
Rufous-sided Towhee	21.25	2.73	1.10	9.77
TOTAL	89.35	29.27	46.83	205.05

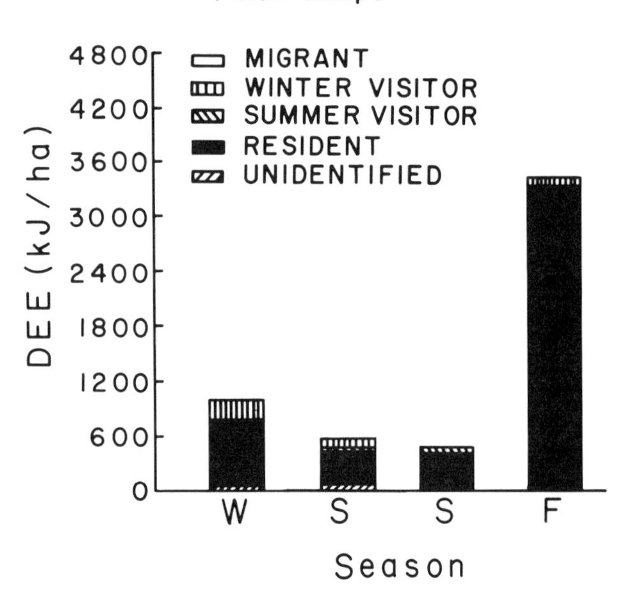

FIGURE 43 The sudden late-September appearance of large flocks of Common Ravens accounts for 78 percent of the fall peak in daily energy expenditure (DEE).

The three most abundant residents of the piñon-juniper, the Mourning Dove, Common Raven, and Scrub Jay, account for 56 percent of the individuals and 85 percent of the DEE by resident species (table 29). As described above, ravens are abundant during only one month of the year. Thus, their impact on the annual energy flow is extremely transient.

Winter Visitors

During winter, residents and winter visitors comprise roughly equal fractions of the total bird density (fig. 42). Winter visitors arrive early, stay late, and hence are significant components of the avifauna from fall through spring. Because the largest and smallest winter visitors differ in mass by only a factor of three, DEE closely parallels density (table 30). The four

TABLE 29 Average Yearly Density and Daily Energy Expenditure (DEE) of Permanent Residents Recorded During 100 Censuses of the Piñon-juniper Habitat, 1979 and 1980

Species	% Occurrence	Density		DEE	
		birds/40 ha	% of total	kJ/ha	% of total
Mourning Dove	60.0	45.80	22.56	233.79	18.59
Common Raven	23.0	38.75	19.09	717.89	57.07
Scrub Jay	84.0	29.81	14.69	112.92	8.98
Mountain Chickadee	42.0	19.88	9.79	21.94	1.74
House Finch	35.0	13.85	6.82	21.94	1.74
Bushtit	18.0	9.33	4.60	5.98	0.48
Common Flicker	29.0	8.93	4.40	52.73	4.19
Chipping Sparrow	20.0	8.49	4.18	9.64	0.77
Rufous-sided Towhee	36.0	8.21	4.04	19.78	1.57
Plain Titmouse	32.0	5.79	2.85	7.27	0.58
Piñon Jay	14.0	4.81	2.37	21.94	1.74
Black-throated Sparrow	11.0	2.80	1.38	3.28	0.26
Bewick's Wren	9.0	1.60	0.79	1.50	0.12
Gambel's Quail	4.0	1.49	0.73	9.59	0.76
Lesser Goldfinch	1.0	0.70	0.34	0.66	0.05
White-throated Swift	4.0	0.50	0.25	1.66	0.13
Total of 10 others[a]	–	2.24	1.10	15.38	1.22
TOTAL		202.98	99.98	1257.89	99.99

[a]Listed in decreasing density: Nuttall's Woodpecker > Scott's Oriole > Red-tailed Hawk > California Thrasher > Rock Dove = Ladder-backed Woodpecker > American Kestrel > Cooper's Hawk > Wrentit = Roadrunner.

most abundant species account for 86 percent of the total DEE by winter visitors. As a group, winter visitors tend to be smaller than residents. Consequently, although they comprise 37 percent of the birds seen during winter (fig. 42), they use only 25 percent of the energy (fig. 43).

The two most abundant species, Western Bluebird and Dark-eyed Junco, account for nearly 60 percent of all winter visitors (table 30). Both species breed in the coniferous forest, and many of the individuals seen in the piñon-juniper during winter may have come from higher up the mountain. This is uncertain, however, and banding studies would be necessary to document local population shifts.

Summer Visitors

Only three species of summer visitors were encountered during spring and summer censuses of the piñon-juniper (table 31). Although they nest in the piñon-juniper, they are not abundant. Thus, they are relatively unimportant to overall bird density or energy flow. A few Phainopeplas

TABLE 30 Average Density and Daily Energy Expenditure (DEE) of Winter Visitors Recorded During 22 Winter Censuses (Dec.–Feb.) of the Piñon-juniper Habitat, 1979 and 1980

Species	% Occurrence	Density		DEE	
		birds/40 ha	% of total	kJ/ha	% of total
Western Bluebird	54.6	49.48	36.51	95.36	39.78
Dark-eyed Junco	45.5	30.70	22.65	43.74	18.25
White-crowned Sparrow	27.3	19.73	14.56	39.84	16.62
Phainopepla	68.2	15.27	11.27	27.25	11.37
Cedar Waxwing	9.1	10.00	7.38	21.97	9.17
Pine Siskin	13.6	6.88	5.08	7.38	3.08
Brewer's Sparrow	9.1	1.64	1.21	1.76	0.73
Red-breasted Nuthatch	9.1	0.91	0.67	0.92	0.38
Purple Finch	4.6	0.91	0.67	1.49	0.62
TOTAL		135.52	100.00	239.71	100.00

TABLE 31 Average Density and Daily Energy Expenditure (DEE) of Summer Visitors Recorded During 52 Spring and Summer Censuses (Mar.–Aug.) of the Piñon-juniper Habitat, 1979 and 1980

Species	% Occurrence	Density		DEE	
		birds/40 ha	% of total	kJ/ha	% of total
Ash-throated Flycatcher	21.2	2.87	56.27	5.79	63.77
Western Tanager	5.8	1.24	24.31	2.56	28.19
Costa's Hummingbird	4.0	0.99	19.41	0.73	8.04
TOTAL		5.10	99.99	9.08	100.00

occur as breeding summer visitors in the piñon-juniper, but none were recorded during the censuses. A striking contrast exists between the piñon-juniper, with its few summer visitors, and the coniferous forest in which summer visitors are dominant.

Migrants

Only two species of migrants were encountered during censuses in the piñon-juniper habitat: the Yellow-rumped Warbler and the Willow Flycatcher. Their combined spring density was a mere 4.92 birds/40 ha. Consequently, migrants appear to play a negligible role in the structure and function of the piñon-juniper community (figs. 42 and 43). Still, at other times, migrants were abundant, as on the afternoon of 12 May 1980 when the piñon trees at Pinyon Flat campground were aswarm with Townsend's, Hermit, Wilson's, Yellow, and Yellow-rumped Warblers. Townsend's Warblers were especially abundant and were accompanied by Western Tanagers, Black-headed Grosbeaks, and Warbling Vireos. The previous day had been unusually cold and windy for this time of year. Snow had fallen in the coniferous forest, and the late spring storm may have pushed the migrants downslope into the piñon-juniper woodland.

ten

CHAPARRAL

The Mediterranean climate (summer drought, winter rain) occurs in California, central Chile, South Africa, southern and western Australia, as well as in the Mediterranean Sea region. Areas in these regions with thin, stony, mineral-deficient soils support a dense scrub dominated by woody sclerophyllous (hard-leaved) evergreens. They experience a common set of selective forces that are inimical to plant growth: violent winds, frequent fires, high temperatures, and less than 750 mm of precipitation annually. These common selective forces result in a remarkable degree of convergence in Mediterranean scrub communities throughout the world.

In California, the Mediterranean scrub is known as chaparral (originally a Basque word for scrub oak, *chabarro*, which was spelled *chaparro* by the Spanish and used by them to designate California's Mediterranean-type scrub). Four distinctive types of chaparral are found in California, with *dry* or *hard chaparral* growing on the western slopes of the Sierra Nevada foothills, the Transverse and Peninsular Ranges (including Deep Canyon), and the coastal side of the Coast Range. California's hard chaparral resembles the matorral of Chile, with close convergence in both plant (Mooney and Dunn 1970) and bird communities (Cody 1973). Consequently, studies of Deep Canyon's chaparral should complement those of similar Mediterranean scrub habitats.

At Deep Canyon, hard chaparral occurs on the steep, north-facing slopes of the Santa Rosa Mountains between 1,220 and 1,965 m elevation and on south-facing slopes about 200 m higher. Deep Canyon's chaparral is less developed than typical California chaparral (Zabriskie 1979). Its shrubs are more widely spaced and stunted, reflecting the dryness of the Santa Rosa Mountains. Rocky outcrops are common, and in openings beneath the plants the soil is coarse and littered with rock fragments (fig. 44).

The dominant perennial plants of Deep Canyon's chaparral, redshank (*Adenostoma sparsifolium*) and desert ceanothus (*Ceanothus greggii*) have somewhat separate distributions. Redshank forms dense, nearly pure stands on granitic soil, whereas ceanothus follows a broad strip of metasedimentary rock up the mountain's north face (Zabriskie 1979).

FIGURE 44 Dry chaparral covers the steep mountain slopes below the coniferous forest. This habitat, dependent on fire for rejuvenation, last burned in 1940.

Yucca, scrub oak (*Quercus turbinella* and *Q. dumosa*), manzanita (*Arctostaphylos glauca* and *A. patula*), sugar bush, and cacti account for about 50 percent of the relative plant cover within the ceanothus stands (fig. 45). These plants constitute a much smaller fraction in the redshank stands (Zabriskie 1979).

FIRE SUCCESSION

Because of the prolonged dry season and dry, resinous nature of the shrubs, the chaparral periodically experiences intense fires that raze the aerial portions of the plants. Although the fires totally destroy the surface vegetation, below ground the plants remain alive. Many Mediterranean scrub plants are well adapted to fire, and in some species (e.g., *Ceanothus*) scorching actually stimulates seed germination. New shoots may sprout from roots and stumps within days following a fire, but most growth follows the next favorable rain (Plumb 1961). The scrub quickly returns to its former state and reaches maturity within ten to twenty years (Mooney and Parsons 1973).

Mediterranean scrubs that have not burned for forty to sixty years become stagnant and have little annual growth. In old scrubs, toxic and hydrophobic substances accumulate in the soil, canopy closure eliminates subshrubs, and diversity decreases until only a few dominant species

FIGURE 45 Relative plant cover reaches its maximum in Deep Canyon's chaparral. Chaparral dominated by desert ceanothus (in bloom in foreground) is more diverse than that consisting of nearly pure stands of redshank. Snowcapped San Jacinto Peak in background.

remain. Fire rejuvenates these old stands by eliminating dead wood and accumulated toxic soil substances. The rapid regrowth of burned shrubs and the invasion of annuals and herbs, formerly shaded out by the thick canopy, favors the bird inhabitants by providing more food.

The last fire to sweep through Deep Canyon's chaparral occurred in 1940 (Zabriskie 1979). Thus, much of the local chaparral is senescent. Plant species diversity is lower here than elsewhere at Deep Canyon (Zabriskie 1979). Consequently, low plant productivity and a correspondingly impoverished avifauna are to be expected. Indeed, in terms of bird density and energy flow, the chaparral resembles the more barren desert scrub community (see figs. 12 and 14). But despite low bird density and energy flow, bird species diversity actually reaches its highest value here. More breeding species (36) occur here than are found in any of Deep Canyon's other habitats. Thus, low plant species diversity and productivity do not necessarily result in low bird species diversity.

BIRD CENSUSES

Two 1.0-km strip transects have been established in the chaparral. One is on the northern slope of Santa Rosa Mountain, between 1,415 and 1,463-m elevation (Section 14; T. 7 S., R. 5 E., Palm Desert Quadrangle).

The other is on a south-facing slope along the Santa Rosa Mountain road (just west of where the road crosses Garnet Queen Creek) at between 1,780 and 1,860-m elevation (Section 20, T. 7 S., R. 5 E., Palm Desert Quadrangle).

Many of the same bird species occur in both the chaparral and piñon-juniper habitats. Although the common species tend to be the same in both habitats, the most abundant bird in the piñon-juniper, the Mourning Dove, was not encountered during the chaparral censuses. Comparing resident species density in both habitats reveals the following habitat preferences:

Prefer Piñon-juniper	*Prefer Chaparral*
Mountain Chickadee	Lesser Goldfinch
House Finch	Bewick's Wren
Common Raven	California Thrasher
Common Flicker	
Plain Titmouse	

A few species, such as the Scrub Jay and Rufous-sided Towhee, were equally abundant in both habitats. In general, these habitat preferences are in agreement with those given by Miller (1951).

Breeding species dominate the chaparral bird community (36 of the 43 species). The breeding bird community in turn is dominated by insectivorous and omnivorous species. Only four of the thirty-six species eat mainly plant material: Mountain Quail, Lesser Goldfinch, House Finch, and Sage Sparrow. They account for only 15.7 percent of the energy flowing through resident chaparral birds (table 32).

Chaparral birds were particularly furtive during summer and fall censuses. This, and the dense chaparral cover, made it difficult to detect them. Consequently, the actual density may have been greater than the observed density. How much greater is impossible to judge. Plotting bird density by season (fig. 46) suggests that the error may not be large; for although birds should be most detectable during the breeding season, density was actually lowest during spring.

Chance sometimes plays a bigger role in studies of community structure than many ecologists care to admit. A good example of this is the Bushtit population density estimate in the chaparral. Based on sixty-six censuses (a reasonably large sample size), the Bushtit was the third most abundant resident in the chaparral (table 32). Yet Bushtits were detected during only 6 percent of the censuses, a much lower detection frequency than for residents with similar densities. The Bushtit's relatively high density estimate resulted from the chance encounter of a few large flocks. If one of the Bushtit flocks had been located elsewhere during a census, the estimated density would have been quite different. This example illustrates not only the role of chance but also the problem of estimating the density of species that flock.

TABLE 32 Average Yearly Density and Daily Energy Expenditure (DEE) of Permanent Residents Recorded During 66 Censuses of the Chaparral Habitat, 1979 and 1980

Species	% Occurrence	Density		DEE	
		birds/40 ha	% of total	kJ/ha	% of total
Scrub Jay	69.7	18.67	21.37	70.72	25.91
Lesser Goldfinch	40.9	15.39	17.61	14.48	5.30
Bushtit	6.1	10.06	11.51	6.45	2.36
Rufous-sided Towhee	39.4	7.51	8.60	18.09	6.63
Bewick's Wren	27.3	5.33	6.10	5.01	1.84
House Finch	16.7	3.52	4.03	5.58	2.04
Common Raven	19.7	3.27	3.74	60.58	22.20
California Thrasher	25.8	3.15	3.61	13.32	4.88
Western Bluebird	10.6	3.03	3.47	5.84	2.14
Mountain Quail	6.1	2.79	3.19	21.53	7.89
Wrentit	18.2	2.55	2.92	3.49	1.28
Black-throated Sparrow	9.1	2.06	2.36	2.41	0.88
Mountain Chickadee	15.2	1.94	2.22	2.14	0.78
Anna's Hummingbird	9.1	1.33	1.52	1.17	0.43
Cooper's Hawk	12.1	1.21	1.38	12.42	4.55
Brown Towhee	10.6	1.09	1.25	2.89	1.06
Sage Sparrow	6.1	0.97	1.11	1.27	0.47
Red-tailed Hawk	10.6	0.85	0.97	18.03	6.61
Plain Titmouse	7.6	0.85	0.97	0.74	0.27
White-throated Swift	6.1	0.60	0.69	1.99	0.73
Total of 6 others[a]	–	1.20	1.37	4.73	1.73
TOTAL		87.37	99.99	272.88	99.98

[a]Listed in decreasing density: Common Flicker > Rock Wren = Brown-headed Cowbird > Band-tailed Pigeon = Poor-will = Costa's Hummingbird.

Residents

Fifty-three percent of all birds encountered during chaparral censuses were residents (fig. 46). This group accounts for 69 percent of the energy flowing through the chaparral bird community (see figs. 4 and 46). The energy flow of this group is greater than its density because five of the ten most abundant residents are large birds with high energy demands: Scrub Jay, Rufous-sided Towhee, Common Raven, California Thrasher, and Mountain Quail. In contrast, the most abundant winter visitors (Dark-eyed Junco and White-crowned Sparrow) are small. Thus, although winter visitors comprise 28 percent of all birds found, they use only 19 percent of the annual energy (see fig. 4).

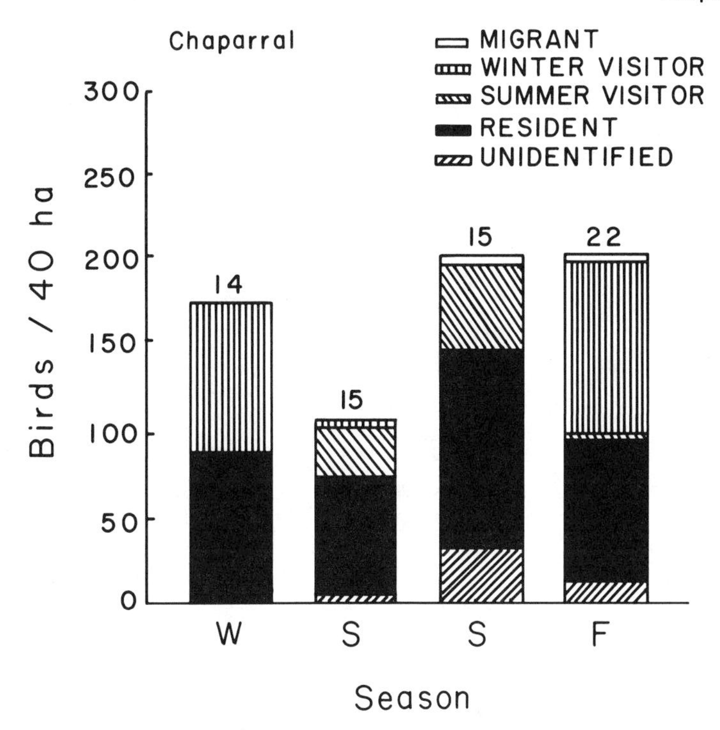

FIGURE 46 Chaparral bird density was lowest in spring. During fall and winter, winter visitors dominated the chaparral community. See fig. 16 for explanation of symbols.

Winter Visitors

During fall and winter, there was such an influx of sparrows that winter visitors became as common as, or more common than, residents (fig. 46). Dark-eyed Juncos constituted the vast majority of the winter visitors (table 33), and their fall arrival coincided with a decline in junco numbers in the adjacent coniferous forest (table 34). These data suggest that juncos might be shuttling between the chaparral and coniferous forest habitats, rather than undertaking the usual long-distance seasonal migration. Because both energy cost and mortality would be lower during a more limited down-mountain migration, this pattern of population shift would have adaptive value. Although it is difficult to imagine how an altitudinal migration pattern could evolve from a latitudinal one, banding studies of Deep Canyon's juncos might nevertheless be well worth the effort.

Summer Visitors

The Black-chinned Sparrow was the most abundant of the ten summer visitor species found during the chaparral censuses. It accounted for nearly 52 percent of all individuals (table 35). Territorial males of this species advertise their readiness to breed by singing from bush tops. Their

TABLE 33 Average Density and Daily Energy Expenditure (DEE) of Winter Visitors Recorded During 14 Winter Censuses (Dec.–Feb.) of the Chaparral Habitat, 1979 and 1980

Species	% Occurrence	Density		DEE	
		birds/40 ha	% of total	kJ/ha	% of total
Dark-eyed Junco	42.9	60.00	68.63	85.48	64.69
White-crowned Sparrow	28.6	13.71	15.68	27.68	20.95
Fox Sparrow	28.6	4.57	5.23	9.64	7.29
Pine Siskin	7.1	4.57	5.23	4.90	3.71
Ruby-crowned Kinglet	14.3	4.00	4.58	3.13	2.37
Golden-crowned Sparrow	7.1	0.57	0.65	1.30	0.99
TOTAL		87.42	100.00	132.13	100.00

TABLE 34 Seasonal Change in Dark-eyed Junco Density at Deep Canyon

	Birds/40 ha			
	Winter	Spring	Summer	Fall
Coniferous forest	2.4	48.7	54.0	26.1
Chaparral	60.0	2.7	0.0	66.6

TABLE 35 Average Density and Daily Energy Expenditure (DEE) of Summer Visitors Recorded During 30 Spring and Summer Censuses (Mar.–Aug.) of the Chaparral Habitat, 1979 and 1980

Species	% Occurrence	Density		DEE	
		birds/40 ha	% of total	kJ/ha	% of total
Black-chinned Sparrow	60.0	21.33	51.92	22.87	38.39
Violet-green Swallow	36.7	8.00	19.47	16.43	27.58
Ash-throated Flycatcher	33.3	5.07	12.34	10.24	17.19
Gray Vireo	16.7	1.60	3.89	1.87	3.13
House Wren	13.3	1.60	3.89	1.59	2.67
Orange-crowned Warbler	10.0	1.07	2.60	1.01	1.69
Scott's Oriole	6.7	1.07	2.60	2.58	4.33
Black-headed Grosbeak	10.0	0.80	1.95	2.15	3.61
Green-tailed Towhee	3.3	0.27	0.66	0.53	0.90
Western Wood Pewee	3.3	0.27	0.66	0.31	0.51
TOTAL		41.08	99.98	59.58	100.00

oft-repeated song is loud and quite distinctive, which makes them easy to detect. Although this undoubtedly had an impact on the density estimate, Black-chinned Sparrows are truly abundant.

Summer visitors utilize 7.8 percent of the annual energy expended by chaparral birds. Their share of the energy pie is highest during the spring, when 17 percent of the estimated community energy flow passes through them (fig. 47).

Migrants

During spring, most migrating birds avoid the chaparral, preferring Deep Canyon's other habitats (especially the desert washes). Wilson's Warbler was the only migrant encountered in the chaparral during spring, and only a few individuals were seen. During fall, many migrating warblers and hummingbirds travel south through California's mountains, and it is then that most migrants appear in the chaparral (table 36).

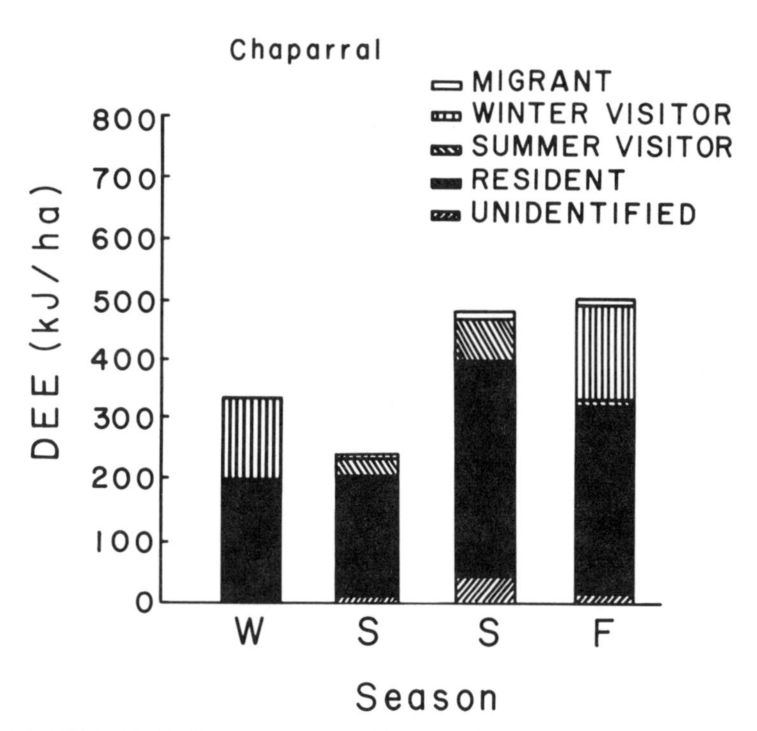

FIGURE 47 Daily energy expenditure of chaparral birds by season. Resident species such as the Scrub Jay, Rufous-sided Towhee, Common Raven, California Thrasher, and Mountain Quail are large and have high energy demands. Thus, residents account for 69 percent of the annual energy flow through this bird community.

TABLE 36 Average Density and Daily Energy Expenditure (DEE) of Fall Migrants Recorded During 22 Fall Censuses (Sept.–Nov.) of the Chaparral Habitat, 1979 and 1980

Species	% Occurrence	Density		DEE	
		birds/40 ha	% of total	kJ/ha	% of total
Rufous Hummingbird	13.6	1.45	33.26	1.20	29.27
Yellow-rumped Warbler	13.6	1.45	33.26	1.65	40.19
Black-throated Gray Warbler	9.1	0.73	16.74	0.66	16.03
Wilson's Warbler	9.1	0.73	16.74	0.59	14.50
TOTAL		4.36	100.00	4.10	99.99

eleven

CONIFEROUS FOREST

A forest dominated by jeffrey pine covers the upper portions of the Deep Canyon Transect, from the edge of the chaparral to the top of Toro Peak. Much of the relatively open forest occupies steep, dry slopes that are liberally punctuated by rocky outcrops (fig. 48). The forest's openness is amplified by low annual precipitation, cattle grazing, and logging. The soil is shallow, and flat areas, such as Stump Spring Meadow, are uncommon. The forest floor is dry in summer but snow-covered during parts of winter.

This type of forest formation, which occurs at middle elevations throughout western North America, consists of relatively open stands of conifers dominated by either ponderosa (*Pinus ponderosa*) or jeffrey pine. Ponderosa pine forests usually occur where conditions are somewhat warmer and moister than the cool, dry situations favored by the jeffrey pine. The nearest ponderosa pines occur to the northwest of Deep Canyon in the San Jacinto Mountains.

Jeffrey pine is one of five cone-bearing plants in Deep Canyon's coniferous forest habitat. White fir (*Abies concolor*) occurs throughout the forest and is usually subdominant to jeffrey pine. But at the highest elevations, it becomes the dominant species and accounts for 43 percent of the relative plant cover (Zabriskie 1979). Limber pine (*Pinus flexilis*) is distributed locally, occupying the final slopes (above 2,500 m) that lead to Toro Peak. Sugar pine (*Pinus lambertiana*), recognized by the elongated cones that dangle from its branch tips, is concentrated on the sheltered slopes of steep ravines. Incense cedar (*Calocedrus decurrens*) occurs throughout the forest, but like sugar pine it prefers moister, more sheltered sites.

The diverse coniferous forest provides a variety of microhabitats for birds. Near the forest's chaparral border, golden-cup oaks (*Quercus chrysolepis*) form dense thickets beneath the open pine canopy. Elsewhere the forest floor supports numerous kinds of shrubs including the genera *Arctostaphylos, Ribes, Amorpha, Symphoricarpos, Gutierrezia,* and *Chrysothamnus.* On sheltered slopes, along streams, and in ravines, the forest becomes

FIGURE 48 An open forest of jeffrey pines dominates the Santa Rosa Mountains' upper slopes. View toward the north shows the Coachella Valley in the background.

denser and more diverse with the addition of several herbaceous species, including sedges, rushes, and monkey-flower. The wide range of microhabitats found at the top of the Deep Canyon Transect supports a variety of bird species.

The coniferous forest's avifauna differs markedly from that of the transect's lower regions. Species such as the White-headed Woodpecker, Pygmy Nuthatch, Steller's Jay, Brown Creeper, and Acorn Woodpecker are restricted to this habitat. Other species, which occur more widely during migration, are confined to the coniferous forest for breeding. These include the Dusky Flycatcher, Gray Flycatcher, Olive-sided Flycatcher, and White-breasted Nuthatch.

BIRD CENSUSES

A single 1 km long strip transect has been established near 2,380 m elevation on the road that connects the top of Santa Rosa Mountain and Toro Peak (Sections 26, 27; T. 7 S., R. 5 E., Palm Desert Quadrangle). This transect passes through generally dry and open forest, and its eastern terminus is marked by Stump Spring Meadow.

The annual cycle of bird density in the coniferous forest follows a distinct and predictable pattern (fig. 49). During winter, the forest is cold

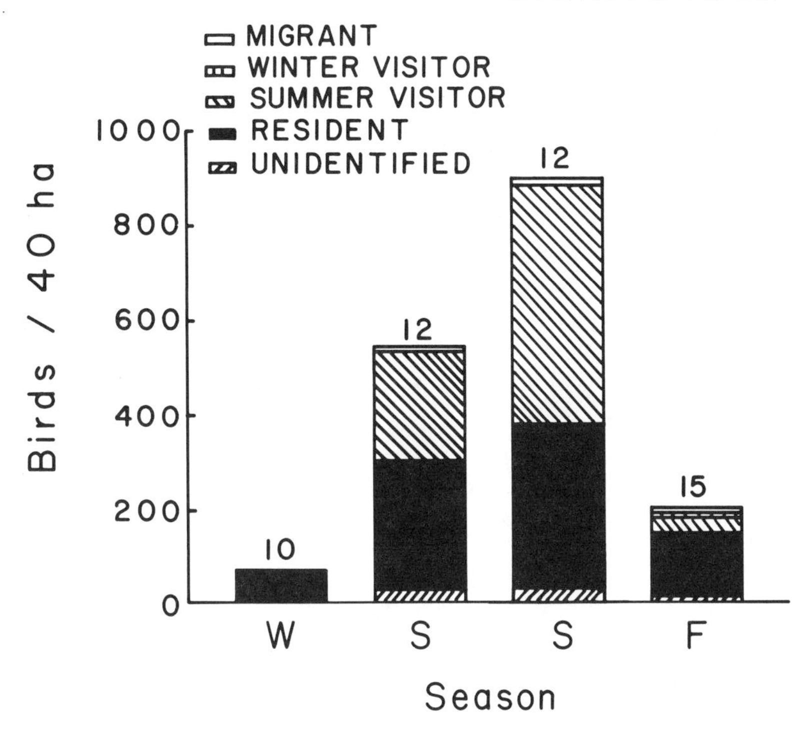

FIGURE 49 In winter, the cold coniferous forest supports only resident species. During summer, a large influx of visitors takes advantage of the milder conditions and breeds. See fig. 16 for explanation of symbols.

and silent, and the ground is often covered with snow. Few species are adapted to such harsh conditions, and bird density reaches its nadir at that time (72 birds/40 ha). With the arrival of spring, conditions begin to moderate, and density increases as birds return to begin another breeding cycle. Density peaks in summer at 890 birds/40 ha, twelve times the winter value. Then it declines precipitously in fall as the summer visitors depart.

During summer, the Pygmy Nuthatch, Mountain Chickadee, and Violet-green Swallow dominate bird density and energy flow in the coniferous forest. They represent only 4 percent of the species but account for 76.7 percent of the individuals (682 out of 890 birds/40 ha) and 64 percent of the daily energy flow (1,156 out of 1,806 kJ/ha). Their proportionately smaller share of the total energy flow reflects their small size. All three species are hole nesters, and they compete intensely for available nest sites. The swallow is the largest of the three (15 grams versus 10 to 11 grams), and its summer density (439 birds/40 ha) is nearly twice the combined density of its smaller competitors (243 birds/40 ha). Whether the swallow's greater density results from some size-related competitive advantage or some other factor (e.g., different method of foraging) is unknown.

TABLE 37 Average Yearly Density and Daily Energy Expenditure (DEE) of Permanent Residents Recorded During 49 Censuses of the Coniferous Forest Habitat, 1979 and 1980

Species	% Occurrence	Density		DEE	
		birds/40 ha	% of total	kJ/ha	% of total
Pygmy Nuthatch	98.0	90.12	43.67	90.80	21.75
Mountain Chickadee	81.6	55.18	26.74	60.91	14.59
Dark-eyed Junco	73.5	33.63	16.30	47.91	11.48
Band-tailed Pigeon	34.7	7.35	3.56	80.73	19.34
Steller's Jay	30.6	3.59	1.74	17.31	4.15
Common Flicker	26.5	3.43	1.66	20.25	4.85
Hairy Woodpecker	26.5	3.27	1.58	11.96	2.86
White-headed Woodpecker	28.6	2.94	1.42	9.75	2.34
White-breasted Nuthatch	24.5	2.29	1.11	3.26	0.78
Common Raven	16.3	2.12	1.03	39.28	9.41
Red-tailed Hawk	10.2	1.63	0.79	34.57	8.28
Brown Creeper	4.1	0.82	0.40	0.71	0.17
TOTAL		206.37	100.00	417.44	100.00

Residents

The resident bird community of the coniferous forest exhibits two characteristics found in stressful environments: low species richness and evenness. Of the 82 species found in the coniferous forest (Appendix I), only 12 (14.6 percent) are permanent residents (table 37). This is the lowest resident species richness of all Deep Canyon's habitats; lower even than that of the hot valley floor, which has 18 resident species. Species evenness is also low, with the three most abundant residents (Pygmy Nuthatch, Mountain Chickadee, and Dark-eyed Junco) accounting for 86.7 percent of all resident birds. The Pygmy Nuthatch alone represents 43.7 percent of all resident birds (table 36). Presumably, both low evenness and richness are related to this habitat's harsh winter climate. But other factors may be involved for, compared with other forests, this one is quite dry and small in area.

The seasonal change in resident bird density (fig. 49) is entirely because of the three most abundant species. The density of the other residents remains stable throughout the year. Mountain Chickadees and Dark-eyed Juncos become more abundant in the coniferous forest in spring and summer, partly because of immigration, as noted earlier (chapters 9 and 10). Some of the summer increase in chickadee and Pygmy Nuthatch numbers, however, probably represents the production of young, as both of these species lay large clutches (6 to 10 eggs).

Winter Visitors

Scattered individuals of a few species, such as Clark's Nutcracker, Williamson's Sapsucker, and the Red-breasted Nuthatch occurred in the coniferous forest during the winter, but they were not detected during the censuses. These birds probably came from the higher slopes of the San Jacinto Mountains.

Summer Visitors

In summer, the coniferous forest's cool shade contrasts strikingly with the valley floor's bright sun and oppressive heat. Accordingly, many species find the coniferous forest an inviting place to breed, whereas only a few are attracted to the valley floor. The forest comes alive in summer, but the valley floor seems desolate and empty, as if waiting for winter.

Despite being present for only half the year, summer visitors are major components of the coniferous forest bird community. They account for nearly 50 percent of the annual energy flow and bird density (fig. 50). During the summer months, summer visitors are the dominant community component, exerting tremendous competitive pressure on the resident species.

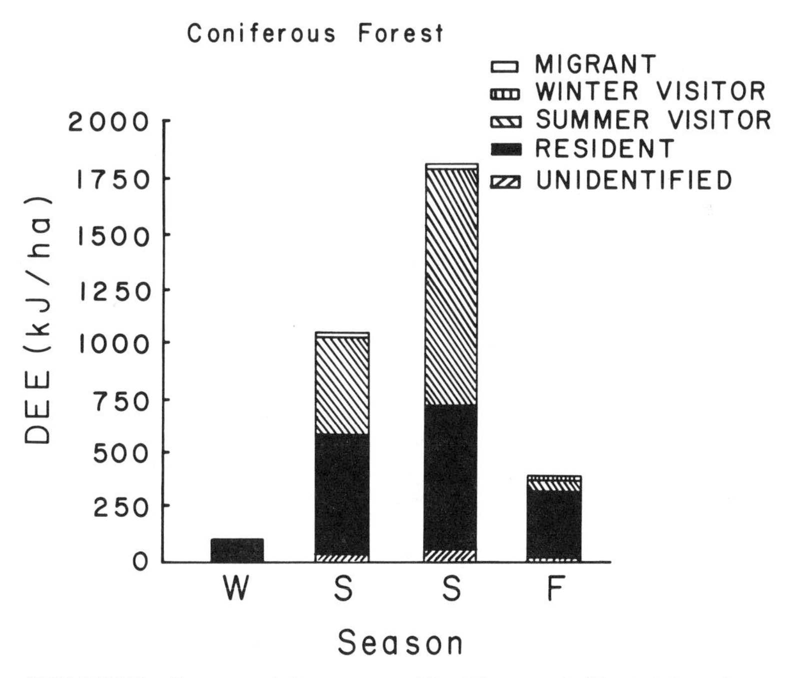

FIGURE 50 Summer visitors account for 47 percent of the total yearly energy flow through the coniferous forest's birds.

Twenty species of summer visitors were recorded during the censuses (table 38). Fifteen additional species of summer visitors were found in the coniferous forest but were not encountered during the censuses. Seven of these were locally distributed along streams, an area not represented on the strip transect. These were Dusky Flycatcher, Western Flycatcher, Solitary Vireo, Warbling Vireo, Northern Oriole, Western Tanager, and Black-headed Grosbeak. Eight others either were associated with oak thickets or were present in low density: Gray Flycatcher, Olive-sided Flycatcher, Tree Swallow, Orange-crowned Warbler, Yellow-rumped Warbler, Black-throated Gray Warbler, Purple Finch, and Lawrence's Goldfinch. Combining these fifteen species with the twenty found during the censuses gives a total of thirty-five summer visitors, a much larger number than found in the other habitats (see table 4).

Although the coniferous forest hosts many summer visitors, the three most abundant species account for 95 percent of all the individuals (table 38). The Violet-green Swallow alone accounts for 82 percent of all

TABLE 38 Average Density and Daily Energy Expenditure (DEE) of Summer Visitors Recorded During 24 Spring and Summer Censuses (Mar.–Aug.) of the Coniferous Forest Habitat, 1979 and 1980

Species	% Occurrence	Density		DEE	
		birds/40 ha	% of total	kJ/ha	% of total
Violet-green Swallow	83.3	303.08	82.25	622.38	82.08
Western Bluebird	83.3	28.34	7.69	54.62	7.20
Cassin's Finch	58.3	17.33	4.70	33.40	4.40
House Wren	33.3	5.67	1.54	5.64	0.74
Western Wood Pewee	33.3	3.67	1.00	4.16	0.55
House Finch	4.2	1.34	0.36	2.12	0.28
Green-tailed Towhee	12.5	1.00	0.27	1.97	0.26
Yellow-rumped Warbler	12.5	1.00	0.27	1.14	0.15
Zone-tailed Hawk	8.3	1.00	0.27	16.99	2.24
American Robin	8.3	1.00	0.27	3.92	0.52
Lesser Goldfinch	8.3	1.00	0.27	0.94	0.12
Ash-throated Flycatcher	8.3	0.67	0.18	1.35	0.18
Mourning Dove	4.2	0.67	0.18	3.42	0.45
Fox Sparrow	4.2	0.67	0.18	1.41	0.19
Total of 6 others[a]	–	2.04	0.55	4.84	0.64
TOTAL		368.48	99.98	758.30	100.00

[a]Each having a density of 0.34 birds/40ha; American Kestrel, White-throated Swift, Say's Phoebe, Chipping Sparrow, Rufous-sided Towhee, Calliope Hummingbird.

summer visitors. Could this high swallow density be a censusing artifact? Swallows are very active aerial insectivores and spend much of the day on the wing. If the same individual, flying back and forth in search of food, were counted more than once during a census, the density estimate would be inflated. Some "double counting" undoubtedly occurred, yet I do not believe it accounts for the high swallow density for the following reason. A nest census conducted on 20 June 1979 near the census transect revealed four swallow nests on 2 ha. This is equivalent to 80 pairs per 40 ha. If each pair fledged an average of 1.8 young, the total swallow density (adults plus young) would equal the observed value, 303 swallows/40 ha. Because Violet-green Swallows lay a single brood of 4 to 5 eggs, they would need a nesting success of only 40 percent to account for the observed density—a reasonable percentage for a hole-nester.

Migrants

Fall migrants were seven times more numerous than spring migrants (1.34 versus 9.60 birds/40 ha). Most migrants are insectivorous, and their relative dearth in spring may result from the presumably low abundance of insects in the coniferous forest at that time of year. Warblers, the most numerous group of migrants, accounted for nearly 90 percent of all individuals seen during August and September (table 39). Overall, migrants were insignificant components of the coniferous forest avifauna, accounting for less than 1 percent of the total density and energy flow (figs. 49 and 50).

TABLE 39 Average Density and Daily Energy Expenditure (DEE) of Fall Migrants Recorded During 17 Fall Censuses (Aug.–Sept.) of the Coniferous Forest Habitat, 1979 and 1980

Species	% Occurrence	Density		DEE	
		birds/40 ha	% of total	kJ/ha	% of total
Wilson's Warbler	23.5	4.24	33.39	3.45	30.63
Orange-crowned Warbler	11.8	2.35	18.50	2.21	19.64
Hermit's Warbler	17.7	1.88	14.80	1.81	16.05
Townsend's Warbler	17.7	1.88	14.80	1.70	15.13
Rufous Hummingbird	11.8	0.94	7.40	0.78	6.90
Hammond's Flycatcher	5.9	0.47	3.70	0.47	4.21
Nashville Warbler	5.9	0.47	3.70	0.41	3.67
Black-throated Gray Warbler	5.9	0.47	3.70	0.42	3.76
TOTAL		12.70	99.99	11.25	99.99

twelve

STREAMSIDE

Deep Canyon's headwaters are the springs and seeps of the upper slopes of the Santa Rosa, Martinez, and Sheep mountains. They begin as trickles, coalesce with the flow from side channels, enlarge, and plunge through the canyon's rocky gorge only to disappear, 16 km from their origin, into the sandy bed of Deep Canyon Wash. The waters are replenished each spring by snow melt, and the vernal flow may be gradual and prolonged, if spring arrives slowly, or swift and torrential, if warm rain melts a heavy snowpack. Stream flow diminishes throughout summer but seldom ceases, and permanent water exists in the canyon bottom as isolated pools.

The comparatively abundant water favors plants that are vegetatively and floristically distinct from those of the adjacent dry slopes and creates a distinct streamside habitat. The vegetative difference between streamside and adjacent habitats is smallest in the coniferous forest, increases downslope, and becomes greatest in the inner gorge near 980-m elevation, where a riparian woodland abuts the dry, desert slopes.

A riparian woodland lies near the junction of Horsethief and Deep Canyon Creeks (fig. 51). There, the canyon becomes broader and flatter and supports a woodland of broad-leaved deciduous trees that rises above an understory of brush and vines. Prominent species include cottonwood (*Populus fremontii*), willow (*Salix* spp.), alder (*Alnus rhombifolia*), and ash (*Fraxinus velutina*). Birds found at Deep Canyon that are known to prefer this habitat for nesting include Cooper's Hawk, Black-chinned Hummingbird, Common Flicker, Tree Swallow, Warbling Vireo, Northern Oriole, Black-headed Grosbeak, and American Goldfinch. Zabriskie (personal communication) found a Cooper's Hawk nest in a cottonwood tree along Horsethief Creek, the only conclusive evidence that any of the above species nest in this woodland. The streamside woodland received only a cursory exploration for breeding birds, however, and the noise of the rushing stream made it difficult to hear bird songs. All of the above species (except the Tree Swallow) nest along streams elsewhere on the Deep Canyon Transect.

FIGURE 51 A riparian woodland dominated by cottonwoods develops near the confluence of Horsethief and Deep Canyon Creeks.

In California, a larger number of bird species (75) is associated with the riparian woodland than any other plant assemblage (Miller 1951). For birds, this woodland is the most luxuriant of the strictly terrestrial assemblages. Its light green foliage and water act as powerful attractants for birds. At Deep Canyon, more bird species (114) were recorded from the streamside than from any other habitat (Appendix I). Miller (1951) lists thirty-three species of birds with a primary affinity for riparian woodlands. Of these, only eight nest in Deep Canyon's streamside habitat: along Garnet Queen Creek—Cooper's Hawk, Warbling Vireo, and American Goldfinch; along Omstott Creek—Brown-headed Cowbird and Black-headed Grosbeak; along Deep Canyon Creek—Cooper's Hawk, Common Flicker, Hooded Oriole, and Northern Oriole.

Groves of a few to over fifty fan palms (*Washingtonia filifera*) occur between 350 to 910-m elevation in Deep Canyon's narrow side channels. Here they find protection from the destructive floods that scour the main gorge. The stately trees, nestled behind natural rock dams, appear as hanging gardens above the canyon floor. These groves are the favorite nesting site of the Hooded Oriole, which uses the palm fibers for nest construction. Fan palms provide shaded and secure nest sites for several other species, including Scott's Orioles and the ubiquitous House Finch.

Bird Censuses

Three 1-km strip transects have been established in the streamside habitat at three different elevations. The highest strip transect, between 1,753 and 1,860-m elevation, follows Garnet Queen Creek west from where the creek crosses the Santa Rosa Peak road (Section 20; T. 7 S., R. 5 E., Palm Desert Quadrangle). The first 200 m of the transect's eastern end lies in coniferous forest. The remainder of the transect, bordered by chaparral, is dominated by arroyo willow (*Salix lasiolepis*), golden-cup oak, false indigo (*Amorpha fruticosa*), and California barberry (*Berberis dictyota*). The middle strip transect, between 1,073 and 1,171-m elevation, begins near Dos Palmas Spring and follows Carrizo Creek westward to where an abandoned road crosses the creek bed (Section 26; T. 6 S., R 5 E., Palm Desert Quadrangle). The dominant plants on this transect include desert willow, sugar bush, juniper, squaw waterweed (*Baccharis segiloides*), cat's claw, false indigo, and desert apricot. The transect borders piñon-juniper habitat. This accounts for many of the dominant plants, as well as the presence of a few scrub oak and piñon trees. The lowest strip transect is located in Deep Canyon Gorge between 317 and 366-m elevation (Sections 19, 20; T. 6 S., R. 6 E., Palm Desert Quadrangle). This transect follows the rocky streambed and is bordered by the rocky slopes habitat. The dominant streambed vegetation is desert willow.

The streamside habitat does not support a distinctive avifauna. None of Deep Canyon's species is restricted to it, and the few species that prefer it for nesting nest elsewhere as well. Generally, the species encountered along the streamside are identical to those of the adjacent habitat. Thus, where the stream passes through chaparral, we find Brown Towhees, California Thrashers, Scrub Jays, Anna's Hummingbirds, Bewick's Wrens, Bushtits, and similar chaparral species. In the rocky slopes habitat, we find such typical species as Loggerhead Shrikes, Say's Phoebes, White-throated Swifts, Cactus Wrens, and Black-throated Sparrows. Although birds presumably use the streamside habitat as a corridor, and thereby extend their altitudinal ranges, this has yet to be quantified at Deep Canyon.

Sixty censuses of the streamside habitat revealed a tendency for bird density to increase with elevation (table 40). This parallels the general trend of increased bird density with elevation found at Deep Canyon (see

TABLE 40 Relation of Mean Bird Density to Season Determined from Strip Transect Censuses of the Streamside Habitat

	Elevation	Birds/40 ha			
	(meters)	Winter	Spring	Summer	Fall
Garnet Queen Creek	1,800	290 (3)[a]	67 (3)	675 (3)	758 (5)
Carrizo Creek	1,100	200 (12)	290 (6)	405 (6)	415 (12)
Deep Creek	350	40 (3)	269 (3)	16 (2)	112 (5)

[a]Values are means of the indicated number (in parentheses) of censuses.

fig. 12). Table 40 also indicates that, at a given elevation, streamside bird density tends to be higher in summer and fall than in winter and spring. This latter finding derives partly from birds being attracted to water during the hottest and driest times of the year.

thirteen

SPECIES ACCOUNTS

Two hundred seventeen species have been sighted on the Deep Canyon Transect. Some of these are rare visitors or transients, seen only a few times, and these are readily apparent from Appendix I. Although such birds are of interest to bird watchers or zoogeographers, because of their rarity they have little impact on other birds or the environment. For this reason, I emphasize the major species in the following accounts.

Each bird family occurring at Deep Canyon is described, and all of the 112 breeding species receive some mention. I focus on terrestrial species because water birds are of widespread occurrence in North America and generally show few regional distinctions, except for relative abundance (Behle 1978). Furthermore, although water birds are natural components of Deep Canyon's avifauna, appearing on the valley floor's ephemeral ponds, most records are of birds attracted to Deep Canyon by golf course water hazards.

The names given to birds change as our knowledge about them increases. Hence, in the following accounts the currently accepted vernacular and scientific names are both given, along with their more common twentieth century synonyms. Providing synonyms will reduce the confusion that may result when reading the older literature. For example, the Dusky Flycatcher (*Empidonax oberholseri*) was formerly called the Wright Flycatcher (*E. wrightii*) (Grinnell and Swarth 1913) with *E. wrightii* now referring to the Gray Flycatcher. Although both the vernacular and scientific names changed in this example, frequently one of the names is fairly stable—usually the scientific species name. I identify birds only to the species level, which reflects my interest in the functional aspects of avian ecology. The species' geographical range is derived from the fifth edition of the Check-list of North American Birds (American Ornithologist's Union 1957).

LOONS—GAVIIDAE

Loons are restricted to the northern hemisphere. They breed on freshwater lakes of the tundra and boreal zones and migrate south to spend the winter on inshore tidal waters and inland lakes and seas. For over 10,000 years loons have been attracted to the lakes and seas that periodically formed in the Salton Sink. Common Loons winter on the Salton Sea from October to May and occasionally wander to the golf course water hazards of the Deep Canyon Transect, where they rest and feed on the abundant mosquito fish (*Gambusia* spp.). Before the creation of these artificial "lakes," loons doubtlessly migrated over Deep Canyon without stopping.

Loons—called divers in Europe—are a very ancient group. Fossils of toothed, loonlike birds are known from the vast inland seas that covered North America 135 million years ago, and proper loons were swimming in Cretaceous lakes and seas 100 million years ago. It is likely that the survival of loons over such vast time periods results from their aquatic life style. Lakes and oceans change slowly and thus provide stable environments favorable to continuance.

All loons are principally fish eaters, seizing their prey in underwater pursuit. They are skilled at diving, hence the European vernacular name, with most dives lasting less than two minutes. Loons are unable to walk on or take off from land, and most species need to "patter" over the water in a lengthy take-off run to get airborne. Therefore, they are unlikely inhabitants of deserts, and it is surprising that three of the world's four species have been recorded on the Deep Canyon Transect. But birds are highly mobile, and a single observer standing in one spot for a long time (say, 20,000 years) would see many of the world's species fly by. In 1976, for example, a Laysan Albatross (*Diomedea immutabilis*) (an open-sea bird that normally stays well away from the coast) was seen near Deep Canyon at Morongo Pass (Dunn and Unitt 1977).

GREBES—PODICEPEDIDAE

The strictly aquatic grebes are adapted for surface diving and underwater swimming. They are distinguished from ducks by their tailless look, thin necks, and pointed bills (except the Pied-billed Grebe). Their large feet, partially webbed with paddlelike lobes, and rearward-positioned legs make them excellent swimmers. Grebes are labored fliers and, like loons, must "patter" over the water in a take-off run to get airborne. Their diet consists of aquatic animals (mainly fish, insects, and crustaceans). The larger grebes eat mostly fish, seizing them with their bills, but the Western Grebe is unique in spearing fish with its slender bill. During the breeding season some of the smaller grebes feed largely on invertebrates. This habit in the Eared Grebe is associated with an upturned bill for skimming insects off the water's surface. An unusual and unexplained behavior found in grebes is their habit of eating their own feathers.

Three of the six grebe species found in western North America have been sighted at Deep Canyon. Of these, only the Horned Grebe was unexpected. Both Eared and Western Grebes are winter visitors on the Salton Sea, and a few individuals are attracted to the transect's golf course water hazards. In winter, Eared Grebes are abundant on inland lakes throughout California and Nevada, and they have probably visited Deep

Canyon's pools for many thousands of years. During the winter, Pied-billed Grebes migrate southward and appear at oases on the southeastern deserts (Grinnell and Miller 1944) and on shallow lowland ponds everywhere. Deep Canyon's golf course water hazards are suitable habitat for Pied-billed Grebes, and a few individuals are seen there throughout the year.

PELICANS—PELECANIDAE

Members of the order Pelecaniformes recorded for the Salton Sea include the White Pelican, Magnificent Frigatebird (*Fregata magnificens*), Brown Booby (*Sula leucogaster*), and Blue-footed Booby (*Sula nebouxii*) (Small 1974). None of these species is known to have landed on golf course water hazards near Deep Canyon, but this will surely happen given enough time. I have seen flocks of up to 100 White Pelicans flying a few hundred meters above the alluvial plain, and they seem likely candidates for future occurrence at the water hazards of the Ironwood Golf Course.

HERONS, EGRETS, AND BITTERNS—ARDEIDAE

These long-legged, long-necked birds with long, pointed bills feed primarily by wading in shallow water and taking fish, crustaceans, and frogs. All of the species recorded from the Deep Canyon Transect nest at the Salton Sea. With breeding populations located nearby and birds moving up and down the Coachella Valley, herons and egrets are sometimes found in incongruous habitats (fig. 52). At Palm Desert's golf courses and ecologically related sewage ponds, herons and egrets congregate to feed on the abundant mosquito fish (*Gambusia* spp.). Birds sometimes concentrate at these artificial feeding stations. For example, the following birds were seen at the Palm Desert sewage treatment plant following an accidental fish kill (11 February 1979): 3 Pied-billed Grebes, 16 Great Egrets, 86 Snowy Egrets, 1 Northern Shoveler, 4 Ruddy Ducks, 20 American Coots, 50 Ring-billed Gulls, and 1 Belted Kingfisher. An adjacent golf course water hazard held 2 Cinnamon Teal, 10 American Wigeons, 4 Lesser Scaup, and 10 American Coots, thus illustrating the old adage that with the application of a little water and fertilizer, the desert can be made to bloom.

SWANS, GEESE, AND DUCKS—ANATIDAE

In years with heavy winter rains, ephemeral lakes and ponds form in the Coachella Valley's low-lying regions (see fig. 16). Waterfowl and shorebirds are attracted to these bodies of water. Thus, they are natural com-

FIGURE 52 Incongruous sights, such as this Great Egret in the desert, occur near the golf course water hazards (photograph courtesy of Jan Zabriskie).

ponents of Deep Canyon's avifauna that predate industrial man's settlement. In former times, waterfowl doubtlessly wintered and bred on Lake Cahuilla, much as they do on today's Salton Sea. Most of the waterfowl sightings for the Deep Canyon Transect are from the Ironwood Golf Course water hazards where one species, the Mallard, regularly breeds.

Mallard. *Anas platyrhynchos*
Males 1,026 grams; females 867 grams

Synonyms: Common Mallard; Greenhead; *Anas boschas.*

Range: Breeds from Alaska and northwestern and southeastern Canada to northern Baja California, southern Texas, Illinois, Ohio, and Virginia. Winters southward to south-central Mexico.

Deep Canyon: Year-round resident on golf course water hazards.

Mallards are the most familiar wild duck found on city ponds. They are very tolerant of people and nest on damp ground in concealing cover. The species name *platyrhynchos* means "flat-beaked": from the Greek *platys*, "broad or flat," and *rhynchos*, "nose or beak."

NEW WORLD VULTURES—CATHARTIDAE

Well-known scavengers, vultures share their feeding habit with Red-tailed Hawks, Ravens, and Golden Eagles, all of which eat carrion to some extent. The occurrence of these potential competitors, combined with the desert's low productivity, contributes to the scarcity of vultures at Deep Canyon. Vulture numbers in the western United States have declined over the past 100 years, partly because of the use of poison baits for vermin control. Today, the Turkey Vulture is the only vulture that soars over the transect, but this was not always so. Grinnell and Swarth (1913) provide a secondhand report of California Condors (*Gymnogyps californianus*) nesting on the cliffs above Snow Creek on San Jacinto Peak's north side. In the evening, as sunset creeps across the Coachella Valley, it is easy to imagine condors soaring over Deep Canyon's alluvial plain.

Turkey Vulture. *Cathartes aura*
1,200 to 1,900 grams

Synonyms: Buzzard; Turkey Buzzard; Red-headed Vulture.

Range: Southern Canada south across the United States through the Greater Antilles and Central and South America. In California, part of the population is resident, part migratory.

Deep Canyon: Scarce winter visitor; common during spring and fall migration.

Turkey Vultures are year-round residents in the Mojave Desert (Miller and Stebbins 1964). They do not nest at Deep Canyon, however, and are seen most often in the spring, especially during March and April, when northward migrants move up the Coachella Valley toward their northern breeding grounds. On 30 March 1979, several groups of vultures totaling twenty-five individuals were seen migrating through the valley. They were traveling by using "thermals" to gain altitude and then gliding downward until another thermal was encountered.

Road-killed jackrabbits (*Lepus californicus*), antelope ground squirrels (*Ammospermophilus leucurus*), and reptiles probably constitute this vulture's major food supply in the Coachella Valley.

EAGLES, HAWKS, AND HARRIERS—ACCIPITRIDAE

With 217 species world-wide, the family Accipitridae includes the vast majority of the diurnal birds of prey. Their predatory habits result in a common set of morphological features including keen eyesight, hooked beaks, and powerful feet with sharp claws (except in the nonpredatory Old World vultures). All predaceous members of the family kill by seizing prey with their feet. Death results as the sharp talons pierce the prey's body or as the toes close with crushing force. Other features common to the family's members include building stick nests, laying eggs with shells that are greenish inside, and squirting their excrement forcibly in a horizontal stream.

Sharp-shinned Hawk. *Accipiter striatus*
Males 90–110 grams; females 150–200 grams

Synonyms: Northern Sharp-shinned Hawk; *Accipiter velox.*

Range: Breeds from northwestern Alaska, northern Canada, and Newfoundland south to California, Arizona, and New Mexico and eastward to South Carolina and Alabama. Winters across North America from roughly 45° north latitude south to Costa Rica.

Deep Canyon: Uncommon migrant.

The three North American bird hawks of the genus *Accipiter* have similar habits; and two, the Cooper's and Sharp-shinned Hawk, are similar in appearance. In the field, the Sharp-shinned Hawk is distinguished by its smaller size (Scrub Jay-size versus the Crow-sized Cooper's Hawk). Although Sharp-shinned Hawks have square-tipped tails (versus the rounded tail of the Cooper's Hawk), this is not a reliable field mark in flight. The Cooper's Hawk soars more frequently than its smaller relative, and this is sometimes an aid to field identification. As in many birds of prey, females are larger than males. Thus, a large female Sharp-shinned Hawk might be confused with a small male Cooper's Hawk, but they do not actually overlap in size.

Sharp-shinned Hawks occur in California mainly as winter visitors. A few pairs breed in the state's northern half and southward in the Transition Life Zone to at least the San Bernardino Mountains (Grinnell and Miller 1944).

Their diet consists mostly of small birds, such as juncos and warblers.

Cooper's Hawk. *Accipiter cooperii*
Males 220–310 grams; females 320–500 grams

Synonyms: Blue-backed Hawk; Mexican Hawk.

Range: Similar to that of the Sharp-shinned Hawk, but more southerly.

Deep Canyon: Uncommon resident throughout the transect. Numbers increase in winter.

This species was first described in 1823 by Charles L. Bonaparte from a specimen shot by William Cooper. Bonaparte, a nephew of the Emperor Napoleon, was one of America's greatest ornithologists, occupying a transitional place between Wilson and Audubon. Gruson (1972) describes Cooper as a modest, self-effacing man who did not publish a great deal but was extraordinarily generous in allowing others to use his notes and specimens.

FIGURE 53 Cooper's Hawks, *Accipiter cooperii*, feed chiefly on small birds.

Cooper's Hawks (fig. 53) feed mainly on small birds and mammals, with a decided preference for robin- to flicker-sized birds. As an adaptation for grasping prey, the hawk's toes are long, slender, and equipped on the underside with "bumps." Band-tailed Pigeons are probably the largest of Deep Canyon's birds normally taken by Cooper's Hawks.

Cooper's Hawks nest in the coniferous forest beginning in May. Grinnell and Swarth (1913) found two nests in golden-cup oaks near Garnet Queen Mine at the edge of the Transition Life Zone. Mayhew (personal communication) found a nest and two fledged young in these oaks in the spring of 1980. Zabriskie (personal communication) found a large stick nest, probably belonging to a Cooper's Hawk, in a cottonwood tree in Horsethief Canyon. Three to five eggs are laid, and incubation begins with the first egg. Thus, the chicks hatch on different days and the larger and stronger chicks commonly devour their smaller nest mates. Consequently, there are usually fewer young fledged than eggs laid.

Throughout winter, Cooper's Hawks are fairly common on the alluvial plain. With the coming of spring, numbers decrease, suggesting that some of the birds are winter visitors from northern breeding grounds.

Red-tailed Hawk. *Buteo jamaicensis*
Males 0.8–1.0 kilograms; females 1.0–1.4 kilograms

Synonyms: Western Red-tailed Hawk; *Buteo borealis.*

Range: Central Alaska and central and eastern Canada south through Mexico and Central America to Panama.

Deep Canyon: Common resident found throughout the transect.

This species' plumage is highly variable, and some races even lack the red tail. In any plumage (except the most melanistic), there is a contrast between the dark tips of the primary flight feathers and their light bases. The dark sides of the head also contrast with the light chest. A detailed description of the plumages is given by Friedmann (1950). In some cases, it is impossible to assign a *Buteo* seen in the field to the correct species. At Deep Canyon, however, Red-tailed Hawks far outnumber other buteos, and this species should be considered first when sighting a broad-winged, soaring hawk. Deep Canyon's Red-tailed Hawks belong to the western race *B. jamaicensis calurus*.

Red-tailed Hawks build bulky stick-nests that measure about one meter wide and 1/2-m high. Both members of the pair participate in nest construction, but the female alone incubates the 2 to 4 eggs. Incubation lasts 28 to 32 days, and another 45 days pass before the young are able to fly. Both adults continue to feed the fledglings for several weeks until the young reach juvenile status. This species is not very particular about where it nests and is tolerant of human disturbance. At Deep Canyon, Red-tailed Hawks have nested in the coniferous forest, on ledges in the canyon gorge, and even in palms on the Ironwood Golf Course.

In March, as the breeding season begins, spectacular midair courtship flights occur over Deep Canyon—birds plummeting toward each other on folded wings. As the diving bird approaches, its mate rolls over on its back, extending sharp talons. Stoops are interspersed with sequences of undulating flights on half-folded wings. Such courtship displays may last for many minutes. Red-tailed Hawks often soar with their legs and talons lowered and held stiffly in place. This behavior, which functions both in courtship and territorial advertisement, serves to warn other birds away from the area.

Red-tailed Hawks are intolerant of other birds of prey and attack those that enter the pair's territory. One April I witnessed a spectacular example of interspecific territoriality, as a pair of red-tails harried and repeatedly stooped on a Golden Eagle crossing their territory. The attack occurred high above Deep Canyon's inner gorge and only ended when the eagle left the area.

The diet of Red-tailed Hawks depends partly upon what is available. In deserts, they feed on snakes and small mammals, such as jackrabbits and ground squirrels. Carrion is utilized to some extent, as red-tails have been seen feeding on road-killed jackrabbits (Miller and Stebbins 1944). Small prey, such as antelope ground squirrels, are usually carried to a perch to be eaten. Large jackrabbits, being too heavy to carry, are dissected and eaten on the spot. Hawks gorge themselves when a large animal is killed and may not need to eat again for several days. In captivity, Red-tailed Hawks can survive up to two to three weeks without food or water (Dobbs, personal communication). During extended fasts the hawks' body temperature gradually declines from 42° to 37.5° C, suggesting that their metabolic rate similarly falls. Reducing energy needs while fasting would help the birds survive periods of food shortage.

Zone-tailed Hawk. *Buteo albonotatus*
Males 628 grams; females 886 grams

Synonym: *Buteo abbreviatus.*

Range: Breeds from northern Baja California, Central Arizona, southwestern New Mexico, and western Texas south through Mexico and Central America to northern South America. Winters through most of the breeding range.

Deep Canyon: Summer visitor.

The biggest surprise of this study was finding a pair of Zone-tailed Hawks breeding in the Santa Rosa Mountains: the first state breeding record for the species. Dr. Wilbur W. Mayhew first recorded the species in the transect's coniferous forest in 1978. The following year, a pair nested unsuccessfully atop a tall sugar pine located in a ravine on the mountain's north slope. Unsuccessful nesting attempts followed in 1980 and 1981. The same nest tree was occupied each year, and the pair reused the previous year's nest.

In flight and appearance, the Zone-tailed Hawk superficially resembles a Turkey Vulture. It forages by soaring with wings held upright in a Turkey Vulture-like "V." This led Willis (1963) to suggest that, by resembling the innocuous vulture, the zone-tail is able to surprise unwary prey.

Golden Eagle. *Aquila chrysaëtos*
Males 3.3 kilograms; females 3–4.5 kilograms

Synonyms: *Aquila canadensis*; American Golden Eagle.

Range: Widely distributed throughout the northern hemisphere. Some populations are migratory.

Deep Canyon: Rare, wide-ranging resident.

Occasionally, one or two of these large, dark brown birds with 2-m wingspans are seen soaring over the Deep Canyon Transect. In flight, eagles flap their wings more often than vultures, and their wings appear longer and less tapered than those of buteos. Adult eagles are uniformly dark, but immature eagles have a broad, white tail band and white "mirrors" in their wings.

The Golden Eagle is the largest and most controversial bird of Deep Canyon, controversial in the sense that misconceptions concerning eagles are common, even among trained biologists. To puncture two misconceptions: Golden Eagles kill their prey with their talons, not by biting with the beak; and they are incapable of carrying loads greater than one-quarter to one-half their mass. Thus, eagles do not carry off children, as sometimes has been claimed. But eagles *are* powerful animals, and they can kill animals larger than themselves. Valid accounts exist of Golden Eagles killing young pronghorns (*Antilocapra americana*) weighing up to 44 kg (97 lbs) (Goodwin 1977). There are less reliable records of single eagles killing wolves and adult deer (see Bent 1937a). Occasionally, two or more eagles cooperate to kill large mammals, as, for example, deer caught in snow drifts. Eagles kill large animals by repeatedly striking them on the back and shoulders, inflicting deep wounds with their inch-long talons. Since they cannot carry off large prey, they consume their kill on the spot, sometimes returning to the carcass the following day.

Eagles eat an astonishing array of animals and are highly opportunistic in their choice of prey. At Deep Canyon, Golden Eagles take mostly jackrabbits, quail, and antelope ground squirrels. Some desert bighorn sheep lambs undoubtedly are killed as well.

Each pair of eagles requires a home range of 50 to 150 km^2. Thus, the Deep Canyon Transect probably supports no more than two pairs of eagles. Most of the time there appears to be only one pair present. Eagles have nested at Deep Canyon, but because the inner gorge is so rugged, a thorough search for nests has not been made. In 1978, two adults and one immature eagle were seen soaring over the canyon. The eagles do not appear to nest successfully every year and in some years may shift the nesting site to an adjacent canyon.

Golden Eagles live for up to twenty years. Like most long-lived birds, they are slow to mature (first breeding when four to five years old) and produce few young. Typically, they fledge a single chick from a two-egg clutch.

Bald Eagle. *Haliaeetus leucocephalus*
Males 3.5–4.5 kilograms; females 4–6 kilograms

Synonym: The name has been unchanged since 1900.

Range: In California, breeds in a few locations in the northern portion of the state. As a winter visitor, it ranges the length of the state but is more common in the north.

Deep Canyon: Dr. Kathleen E. Franzreb and Sidney England saw a single Bald Eagle flying over the valley floor in March 1980.

An influx of Bald Eagles into California from more northern breeding sites occurs each winter. Although most sightings of wintering eagles occur in the northern portions of the state, individual eagles range widely over all sorts of terrain. Those that happen to wander onto the Deep Canyon Transect probably belong to the endangered southern race, *Haliaeetus l. leucocephalus.* Although its numbers are now greatly reduced because of man's direct and indirect interference, in the mid-1880s the Bald Eagle was one of the most abundant birds of prey in Washington State. It was also an abundant California resident, particularly along the sea coast (Belding 1890).

Northern Harrier. *Circus cyaneus*
Males 300–400 grams; females 350–550 grams

Synonyms: Marsh Hawk; American Marsh Hawk; *Circus hudsonius.*

Range: In North America, breeds from northern and western Alaska across Canada to Newfoundland, thence southward to Mexico in the west, more northward in the central and eastern United States. The range shifts southward in winter, with some birds reaching Colombia.

Deep Canyon: Uncommon winter visitor and/or migrant of the lower elevations.

Northern Harriers are the ecological counterparts of the partly diurnal Short-eared Owl. Both species forage by coursing low over the ground, surprising small animals such as mice, birds, lizards, and insects. Like the owl, the hawk depends partly on hearing to locate prey. Consequently, the hawk's facial feathers have become modified to form an owllike facial disk, which increases auditory localization. These hawks are capable of taking birds the size of quail (Miller and Stebbins 1964), and Gambel's Quail are a likely diet item. Most harriers encountered at Deep Canyon are passing through on migration. But increased agricultural activity in the Coachella Valley has probably increased the number that linger through the winter and spring months.

CARACARAS AND FALCONS—FALCONIDAE

Falcons are widely regarded for their spectacular flying ability and keen vision. The large North American falcons hunt by flying high above their intended prey and then making a high-speed stoop with nearly folded wings. The speeds reached in the stoop are among the fastest known in the bird world. From studies of a Lugger Falcon (*Falco jugger*) flying in a wind tunnel, Tucker and Parrott (1970) estimated that this falcon's maximal terminal velocity might be 100 m s^{-1} (216 mi h^{-1}). Mebs (1972) observed Peregrine Falcons diving on urban pigeons at speeds of up to 90 m s^{-1} (194 mi h^{-1}).

The falcon's visual power is not as great as is generally supposed. Based on the number of lines discriminated per unit distance, it is about 2.5 times that of man's (Fox et al. 1976). The falcon's main visual advantage over man concerns image focusing. Whereas man's sharpest vision is with the fovea (try reading this while focusing on the book's spine), the image falling on the falcon's retina is in focus at all points (see Sillman 1973). Because our center of sharpest focus is tiny, we must scan the sky to locate a distant pigeon. The falcon sees it at once. This is a tremendous advantage for an aerial predator.

Prairie Falcon. *Falco mexicanus*
Males 450–625 grams; females 750–1,050 grams

Synonyms: None since 1900.

Range: Breeds from central British Columbia, southern Alberta, southern Saskatchewan, and North Dakota south to northern Mexico and east to Texas. Winters from the northern part of the breeding range southward into Mexico.

Deep Canyon: Uncommon resident.

Long, pointed wings (wingspan 1 m), a fairly long tail, and rapid, powerful wingbeats interspersed with glides are the marks of the Prairie Falcon. The adult's underside appears pale buffy, streaked with dark brown. In flight, the axillars and wing lining show black. As its name implies, the Prairie Falcon is a bird of open spaces.

For a falcon, this species is relatively common in the southwestern deserts. On the Deep Canyon Transect, it is most commonly encountered within the inner gorge, on the rocky slopes, or flying across the valley floor and alluvial plain. Prairie Falcons are occasionally encountered perched atop power poles in the Coachella Valley. Some individual falcons are remarkably unwary and do not fly from the power poles when a car stops close by, a trait likely to contribute to death by shooting.

Prairie Falcons require sheer cliff faces for nesting. They prefer to nest in holes but readily use abandoned raven nests if available. Two active aeries are known on the Deep Canyon Transect, but the number of nesting pairs could be higher. Prairie Falcons nest in April and May, and three to six eggs are usually laid.

This desert falcon meets its water and energy needs with prey items such as small rabbits, ground squirrels, and birds. Prairie Falcons use a variety of hunting techniques to capture prey, including soaring over open terrain and watching for passing birds while perched. I have seen Prairie Falcons soar over Deep Canyon's alluvial plain for several minutes without flapping their wings. They apparently were using rising "thermals" of hot air, since they soared in circles. Thermal soaring is typical of broad-winged raptors, but obviously pointed-winged falcons can fly this way as well.

Peregrine Falcon. *Falco peregrinus*
Males 600–750 grams; females 850–1,200 grams

Synonym: Duck Hawk.

Range: Nearly cosmopolitan, but absent from the eastern Pacific islands and New Zealand. In North America, breeds across the arctic south to Baja California and the southern United States. Birds move south in winter, some as far as central Argentina and Uruguay.

Deep Canyon: One occurrence in February.

Declining North American numbers of this falcon have been attributed to pesticides and loss of habitat. Inland populations have fared better than those along the coasts, with populations in California's Sierra Nevada Mountains remaining at near historical levels. In Arizona, Peregrine Falcons have actually increased as a nesting species since 1939 (Phillips et al. 1964).

Merlin. *Falco columbarius*
Males 140–190 grams; females 150–225 grams

Synonym: Pigeon Hawk.

Range: Breeds in North America from the Arctic Ocean south to about 43° north latitude (further south in the mountains). Winters from the southern United States south to northern South America.

Deep Canyon: Rare fall and spring migrant.

Merlins, or Pigeon Hawks as they were formerly called, are capable of killing pigeon-sized birds. But they generally prefer smaller species, such as Dark-eyed Juncos, House Finches, or Lesser Goldfinches. Up to 95 percent of this open-country species' diet consists of birds. Immature Merlins, like many birds of prey, take a wider variety of prey items than adults until their hunting ability is crystallized (Mallette and Gould 1976). A few Merlins may breed in northern California. The species' status in southern California is mainly one of migrant or winter visitor.

American Kestrel. *Falco sparverius*
Males 90–130 grams; females 120–170 grams

Synonyms: Sparrow Hawk; *Cerchneis sparverius.*

Range: Breeding range extends throughout North America from the arctic to Mexico. Northern populations move south in winter.

Deep Canyon: Common resident and transient.

This is both the smallest and most common Californian member of the order Falconiformes. Small size carries with it both benefits and liabilities. (For a lucid discussion of the consequences of size, see Calder 1974.) One liability shared by many small birds is a short life expectancy. Approximately 60 percent of the immature and 46 percent of the adult American Kestrels die each year (Henny 1972). This high mortality rate must be counterbalanced by a high fecundity rate, if the population is to remain stable. Typically, a single clutch of four to five eggs (rarely seven) is produced each season. Hence, demographically, kestrels more closely resemble songbirds than large birds of prey, which typically live longer and produce two to four eggs each season. Kestrels prefer woodpecker holes or natural holes and cavities in trees as nesting sites, but sometimes they use cavities in buildings or rocky banks. Kestrels nest throughout the Deep Canyon Transect in suitable sites. In the coniferous forest, they nest in holes in jeffrey pines excavated by Common Flickers.

Like the Northern Harrier and Merlin, the American Kestrel is sexually dimorphic. Compared with the browner female, the smaller male's wing is mostly blue and his tail rufous, down to the black subterminal band. Most hawks have streaked young. Juvenile kestrels, however, resemble the adults. Furthermore, kestrels are able to breed when one year old, whereas most birds of prey require two to five years to reach sexual maturity.

American Kestrels are opportunistic in their choice of prey. In the Coachella Valley, they prey heavily on large, ground-dwelling insects and lizards. A smaller fraction of their diet consists of small birds and mammals. Man's development of the valley benefits kestrels by increasing prey densities and providing perch sites, such as power lines, from which the birds can hunt. An alternative to the usual sit-and-wait foraging tactic involves hovering at a height of about 10 m while carefully scanning the ground below for prey. This method frequently is employed when winds provide some lift or when hunting open fields that lack perches.

PHEASANTS AND QUAIL—PHASIANIDAE

The family Phasianidae is mainly an Old World group of about 177 species. It includes quail, pheasants, partridges, peafowl, and jungle fowl and is by far the largest group of gallinaceous birds (order Galliformes). Galliformes is an old assemblage of mostly ground-dwelling birds (fossils date from the Eocene 50 to 60 million years ago). Most species have large breast muscles that enable them to "explode" from cover in fast flight. The flight muscles, however, are composed mainly of white fibers, which are rich in fat but lack the mitochondria and abundant vascular supply necessary for sustained aerobic flight. Consequently, flying birds tire quickly, and only a few species are truly migratory (some members of the Old World tribe Coturnicini). The leg muscles, in contrast, are composed of red fibers. Thus, the birds are capable of sustained running. Indeed, when threatened, many species prefer to run rather than fly. Most species have strong feet and typically expose their food by scratching. The group's fondness for the ground even extends to a preference for bathing in dust or sand rather than water.

Outside of the breeding season, quail are found in coveys composed of one or more family groups. At the beginning of the nesting cycle, they separate into pairs. Males without mates isolate themselves, establish calling posts, and are heard repeatedly giving the single *kaah* note or cock call. The nests of Deep Canyon's three quail species are all shallow scrapes on the ground lined with plant material. They usually are placed at the base of a dense bush. In Gambel's and California Quail, eggs are laid at a little more than a 24-hour interval—five a week—until the clutch of 12 to 14 is completed. Incubation begins with the laying of the last egg and lasts 21 to 23 days (Harrison 1978). Usually the female alone incubates, but the male may take over if the female dies. In Mountain Quail, the male has a brood patch and may incubate one of two separate clutches laid by his mate (Harrison 1978). At hatching, young quail are precocial and downy and begin following the adults immediately. In favorable years, the male may care for the young while the female nests again.

Mortality among young quail seems to be high despite constant parental care. Quail have evolved several mechanisms to counter high predation losses, including rapid growth and the use of distraction displays by the adults to lure predators away from the young. The wing feathers develop especially quickly and young quail are capable of short flights within about one week of hatching. I once tried to catch some very young Gambel's Quail and received a lesson in distraction displays. As I chased the chicks, they scattered in several directions, but most took refuge in low bushes. The female initially flushed into a large palo verde tree about 10 m away but quickly returned in response to the young's distress notes. She surprised me by doing a "broken wing" distraction display similar to that of the Killdeer. Fascinated by her behavior, I followed as she led me away from her young.

California Quail. *Lophortyx californicus*
Males 172 grams; females 182 grams

Synonyms: California Partridge; Valley Partridge; Valley Quail; *Callipepla californicus.*

Range: From southern Oregon and western Nevada south to the Cape region of Baja California.

Deep Canyon: Resident of the lower plateau, chaparral, and piñon-juniper habitats.

Much can be learned about the factors that control animal distribution by studying populations along the limits of a species' range. The Santa Rosa Mountains formerly marked the southeasternmost occurrence of California Quail. Thus, Deep Canyon's populations occurred along the edge of the species' range. Unfortunately, quail have been indiscriminately transplanted throughout California on the behalf of sportsmen. Consequently, the gene pool has been vigorously stirred and the original subspecies characteristics hopelessly clouded (Grinnell and Miller 1944).

At Deep Canyon, the California Quail's distribution overlaps that of Gambel's Quail. The two species hybridize in the Pinyon Flat area, and I have seen mixed coveys containing hybrids near Omstott Creek. The first reported hybridization of these species in the wild was based on a specimen collected on 20 November 1903 in Morongo Pass, about 50 km northwest of Deep Canyon (see Miller and Stebbins 1964).

Reproduction in California Quail (fig. 54) is irregular in the more arid portions of its range and depends upon the amount of winter rainfall preceding the spring breeding season. When rainfall is abundant, a rich carpet of forbs is produced, and the birds breed vigorously. In dry years, the ground is sparsely covered with stunted forbs and grasses, and breeding is poor. Leopold and his coworkers (1976) found that changes in the plants' chemical composition were responsible for the variations in breeding success. In dry years, the leaves of the stunted desert annuals accumulate phytoestrogens that, when eaten by the quail, inhibit reproduction (a sort of nature's birth control "pill"). In this way, the production of young is prevented during years of inadequate food. In wet years, the forbs grow vigorously and the phytoestrogens are largely absent.

PLATE 9 Burrowing Owl *Athene cunicularia*

PLATE 10 Roadrunner *Geococcyx californianus*

PLATE 11 Crissal Thrasher *Toxostoma dorsale*

PLATE 12
Ash-throated Flycatcher
Myiarchus cinerascens

PLATE 13 Black-chinned Sparrow *Spizella atrogularis*

PLATE 14 Black-throated Sparrow *Amphispiza bilineata*

PLATE 15 Loggerhead Shrike (*Juvenile*) *Lanius ludovicianus*

PLATE 16 Mourning Dove *Zenaida macroura*

PLATE 17 Costa's Hummingbird *Calypte costae*

PLATE 18 Verdin *Auriparus flaviceps*

PLATE 19 American Kestrel *Falco sparverius*

PLATE 20
Black-tailed Gnatcatcher
Polioptila melanura

PLATE 21 MacGillivray's Warbler *Oporonis tolmiei*

FIGURE 54 Male California Quail, *Lophortyx californicus*. This species hybridizes with Gambel's Quail where their ranges overlap, as, for example, in Deep Canyon's piñon-juniper habitat.

Gambel's Quail. *Lophortyx gambelii*
Males 163 grams; females 178 grams

Synonyms: Desert Quail; Desert Partridge; *Callipepla gambeli.*

Range: Across southern Nevada to western Colorado south into western Texas and northern Mexico.

Deep Canyon: Resident from the valley floor up to the coniferous forest's lower edge.

This bird was named for William Gambel, a nineteenth-century surgeon, naturalist, and adventurer who died in 1849 (at around age thirty) while on a collecting trip to California (see Gruson 1972). Nineteenth-century ornithology was frequently hazardous in North America. On his last trip to California, Gambel managed to survive numerous attacks by Pawnees, only to become trapped by snows in the Sierra Nevada Mountains. He and four other men escaped on foot to the relative security of Rose's Bar on the Feather River, where Gambel immediately contracted typhoid and died a few days later.

Although the ranges of Gambel's Quail and California Quail overlap at Deep Canyon, the latter is restricted to the higher and more mesic habitats, whereas Gambel's Quail is a true desert species. This difference in distribution reflects differing physiological tolerances to desert conditions. Compared with California Quail, Gambel's Quail produce a more concentrated urine, maintain body weight on higher salinities of drinking water, dehydrate more slowly when deprived of drinking water, and decrease (versus increase) activity during heat stress (Carey and Morton 1971). Each of these abilities contributes to the desert success of Gambel's Quail and helps explain the differing habitat preference of these two species.

Gambel's Quail is mainly granivorous from summer through winter. It is somewhat omnivorous during spring when succulent vegetation (especially the green tips of leaves) and insects are sought out. I have frequently seen quail pulling off and crushing chuparosa blossoms to obtain the plant's nectar and ovaries and eating berries of the desert mistletoe. These foods contain appreciable quantities of water and help to make the birds less dependent upon drinking water during the spring.

The breeding season varies somewhat from year to year, depending upon environmental conditions. My wife saw young in mid-March near the Ironwood Golf Course, and Grinnell and Swarth (1913) reported finding a nest with eggs on 30 August. Such extreme dates suggest that two, or possibly three, broods may be raised in a single year. As in the case of California Quail (see above), variations in rainful probably act through plant hormones to regulate breeding in the desert quail.

On the alluvial plain, Gambel's Quail roost at night in large, dense palo verde trees. Certain trees seem to be favored and may contain scores of birds. During the warm, dry months, quail leave their nighttime roosts at first light and travel by foot to water. The journey is usually a leisurely one, with the birds foraging along the way.

Mountain Quail. *Oreortyx picta*
Males 235 grams; females 220 grams

Synonyms: Plumed Quail; Plumed Partridge; Mountain Partridge; *Callipepla picta.*

Range: From southern Washington and southwestern Idaho south to northern Baja California.

Deep Canyon: Resident chiefly from the coniferous forest downmountain to the lower plateau.

Hiking the upper slopes of the Deep Canyon Transect, one is certain to encounter the Mountain Quail. Its loud, clear whistle—a mellow *too wook*—sounds across the hillsides throughout the year. From late summer through winter, Mountain Quail travel in loose flocks, sometimes accompanied by Gambel's Quail. Unlike *Lophortyx* quail, Mountain Quail coveys seldom contain more than twenty birds (Miller and Stebbins 1964). Mountain Quail also produce fewer young than *Lophortyx* quail (a single brood of seven to ten versus two broods of twelve to fourteen). This may be related to differential rates of mortality. Large birds are usually longer-lived than small birds and may suffer less predation. Because Mountain Quail are about one-third larger than *Lophortyx* quail, they probably have a lower overall morality rate and can balance losses with a lower reproductive rate.

During spring, the coveys dissolve into pairs, and laying commences in late April. On 24 May 1979, I encountered a female Mountain Quail and twelve recently hatched chicks beside Deep Canyon Creek below Sheep Overlook (elevation 975 m). The chicks were too small to have traveled far and clearly hatched nearby. This indicates that, at Deep Canyon, Mounain Quail breed from the coniferous forest downslope to the edge of the lower plateau. In the Mojave Desert, Miller and Stebbins (1964) always found flightless young Mountain Quail within half a mile of water. This species generally seems to be very dependent upon surface water for nesting.

Mountain Quail seem to be most abundant in Deep Canyon's chaparral. But the strip-transect method may underestimate their true numbers in dense vegetation, as the birds are wary and shy away when approached.

RAILS—RALLIDAE

Rails are marsh birds. They did not occur at Deep Canyon before the creation of artificial ponds and marshes, and to date, the only rail recorded from the transect is the American Coot, or Mud-hen. Coots are fairly regular winter visitors to sewage ponds and golf course water hazards. Three other rail species that occur at the Salton Sea migrate through the Coachella Valley and might eventually appear on the transect. These are the Virginia Rail (*Rallus limicola*), Sora Rail (*Porzana carolina*), and Common Gallinule (*Gallinula chloropus*).

PLOVERS—CHARADRIIDAE

The order Charadriiformes (shorebirds) contains 328 species, of which sixty-three are plovers (family Charadriidae). Plovers occur throughout the world (except Antarctica), and the smaller "sandplovers" of the genus *Charadrius* are ubiquitous on marine or inland shores. A few species have adapted to open grasslands and deserts. North America's one desert plover, the Killdeer, regularly nests at Deep Canyon.

Killdeer. *Charadrius vociferus*
85 grams

Synonyms: Killdeer Plover; *Aegialitis vocifera.*

Range: Whole of temperate North America. Breeds throughout its

FIGURE 55 Killdeer nest on barren, sandy flats adjacent to golf courses and feed on the fairways. They are the most desert-adapted of North American plovers.

range and winters from California and the Gulf Coast south to northern South America.

Deep Canyon: Permanent resident of lower elevations.

Throughout its range, the Killdeer breeds in open spaces, usually on short turf or bare areas of sand or gravel. Breeding starts in mid-March, with the first young hatching around mid-April. The cryptically colored eggs (usually four) are deposited in a shallow scrape on the ground and are incubated by both parents (fig. 55). At Deep Canyon, Killdeer usually nest near golf courses, which provide ideal foraging habitat and abundant water. Killdeer feed exclusively on small animals such as insects, crustaceans, and other arthropods. These birds run in short bursts and dart at their prey. Parental care of the young is well developed, and the adults often use injury-feigning displays to distract intruders from the nest site. The parents do not feed the chicks.

SANDPIPERS—SCOLOPACIDAE

This family contains eighty-two species of probing shorebirds, of which nine are known to occur at Deep Canyon (Appendix I). None of the species recorded from the transect nest there, however. Deep Canyon's sandpipers merely pass through on migration, attracted to the area by golf course water hazards and the ephemeral ponds that form on the valley floor following heavy rains. Most sandpipers stay in the area only a few days, but a few winter in the area, especially at the Salton Sea. Many of the family's species winter in Central or South America and breed in the far north. Their spring migration is timed so that they arrive on the breeding grounds as the snows begin to melt. Thus, spring migrants pass through the Coachella Valley mainly in mid-March through mid-May, and returning birds pass through in August and September. Cogswell's (1977) detailed accounts of the distribution and occurrence of the Scolopacidae in California should be consulted for more information.

AVOCETS AND STILTS—RECURVIROSTRIDAE

Stilts and avocets—proportionately the longest-legged waders—typically inhabit the shores of shallow alkaline lakes, pans, and lagoons. They eat small aquatic animals taken from the surface of water or submerged mud. The two species seen at Deep Canyon, the American Avocet and the Black-necked Stilt, occur year-round at the Salton Sea. Whether the Deep Canyon birds are stragglers from the Salton Sea or postbreeding migrants from the north is uncertain.

PHALAROPES—PHALAROPODIDAE

Phalaropes are small aquatic wading birds that superficially resemble sandpipers. Only three species exist worldwide, and all three pass through California en route to their northern breeding grounds. The Red Phalarope (*Phalaropus fulicarius*), the most pelagic of the three, is rarely found inland. The other two, Wilson's and Northern Phalaropes, are often seen traversing the Coachella Valley from mid-April through May. The Northern Phalarope is the more abundant of the two, and many were seen around golf course water hazards in May 1980. One even appeared on Deep Canyon Creek near the Boyd Research Center. The fall southern migration occurs chiefly between July and September. At that time, Northern and Wilson's Phalaropes become abundant on the Salton Sea and can be expected on the Deep Canyon Transect.

GULLS AND TERNS—LARIDAE

Gulls and terns are familiar birds of estuaries, bays, and the open sea coast. Some species regularly wander inland, and they probably occur much more frequently on the Deep Canyon Transect than suggested by Appendix I. Since I spent little time around the golf courses, garbage dumps, and sewage plants that attract these birds, I may have missed some species. I occasionally saw gulls in the distance while censusing the valley floor transect. Most appeared to be Ring-billed Gulls, and this species is certainly much more common than indicated by Appendix I.

Two other gulls, the Western and Bonaparte's Gull, warrant special mention. Inland, Western Gulls are decidely uncommon to rare and it is surprising that one was seen at Deep Canyon. Although it is uncertain, the one Western Gull seen in September was probably a yellow-legged bird from the Gulf of California. The yellow-legged race (considered by some to be a separate species, *Larus livens*) breeds in the Gulf, and some birds move northward to the Salton Sea in late June to September. Bonaparte's Gull is the smallest gull normally occurring in California. It superficially resembles terns and is endowed with great powers of flight. In April and May, migrant flocks of Bonaparte's Gulls appear along various major valleys in California (Cogswell 1977). A few fly though the Coachella Valley, and some of these appear at golf course water hazards.

PIGEONS AND DOVES—COLUMBIDAE

A successful and widespread family containing 255 species worldwide, Columbidae occur in a wide variety of habitats, including several of the world's deserts. But they are absent from the polar and subpolar regions. Five of Deep Canyon's six species are permanent residents, including the introduced Spotted Dove. One species, the White-winged Dove, is a summer visitor.

The Columbidae take a variety of foods including berries, nuts, acorns, seeds, buds, and leaves. They have a manner of drinking that is unusual among birds. The bill is immersed and then, instead of lifting the head and trickling the water down the throat, they suck. Most pigeons and doves share a common pattern of reproduction. They all build a simple, platformlike nest of sticks and twigs. Typical nests are extremely flimsy and may be placed virtually anywhere, including the ground. Usually two pale, unmarked eggs are laid, and both parents share in incubation. The helpless, altricial young are initially fed on "pigeon's milk," a curdlike substance secreted by special cells lining both parents' crops. The young birds feed by inserting their broad, soft bill deeply into one parent's throat and eagerly scooping up the regurgitated material. Young columbids grow very rapidly. Ground Dove young are said to be able to fly when eleven days old (Harrison 1978). Re-nesting is the rule: in good years Mourning Doves may lay five clutches.

Band-tailed Pigeon. *Columba fasciata*
300–400 grams

Synonyms: Wild Pigeon; *Chloraenas fasciata.*

Range: From British Columbia, Utah, and north-central Colorado south to Nicaragua.

Deep Canyon: Permanent resident of the coniferous forest.

Excessive shooting in the early twentieth century reduced the number of Band-tailed Pigeons to a point of threatened extinction (Grinnell and Miller 1944). But with legal protection in 1913, their numbers rapidly increased. Unlike most Columbidae, the Band-tailed Pigeon lays but one egg. Thus, it is less able to compensate for hunting losses than prolific species like the Mourning Dove.

Spend a day in the transect's coniferous forest, and you should see the Band-tailed Pigeon. During the nonbreeding season, when it is gregarious, small flocks roost in the tops of dead jeffrey pines. Although conspicuous when it perches and flies high, the Band-tailed Pigeon is one of the forest's less common species. When its preferred diet of acorns and pine nuts is unavailable, it feeds on manzanita fruits, berries of various species, and leaf buds (Bent 1932). Band-tailed Pigeons also feed on piñon nuts and may occasionally descend the mountain slopes to Deep Canyon's piñon-juniper woodland, although they have yet to be recorded there. At Deep Canyon, Band-tailed Pigeons are preyed upon by hunters and Cooper's Hawks but by few other enemies.

White-winged Dove. *Zenaida asiatica*
130–170 grams

Synonym: *Melopelia leucoptera.*

Range: Breeds from southwestern United States south through south-central Mexico. Winters over the breeding range, but uncommon north of Mexico.

Deep Canyon: Summer visitor of the lower elevations.

This uncommon summer visitor reaches the northernmost extent of its breeding range at Deep Canyon. During spring and summer, its call—"*Who cooks for you?*"—is occasionally heard in the urban habitats. The White-winged Dove seems to prefer cities locally, but elsewhere it is equally at home in desert mesquite thickets.

Mourning Dove. *Zenaida macroura*
120 grams

Synonyms: Western Mourning Dove; Turtle Dove; *Zenaidura macroura.*

Range: From Alaska and southern Canada south through Mexico to western Panama. Migrant in the northern part of the range.

Deep Canyon: Permanent resident.

The Mourning Dove is very gregarious and, when not breeding, occurs in flocks. It breeds throughout the transect, most commonly in the Lower and Upper Sonoran Life Zones, and is one of Deep Canyon's most abundant species (fig. 56). The extremely long breeding season begins in December or February and lasts until fall. The laying of two or three clutches per year is average, but exceptional females will lay five.

Mourning Doves find a wide variety of nest sites suitable, including the ground (fig. 57). Nests are frequently exposed to direct sunlight during part of the day. This increases the risk of overheating for incubating birds and their eggs and chicks. On hot days, adults shade the eggs by standing over them with drooping wings. But when air temperature exceeds approximately 44° C (111° F), shading alone is insufficient to

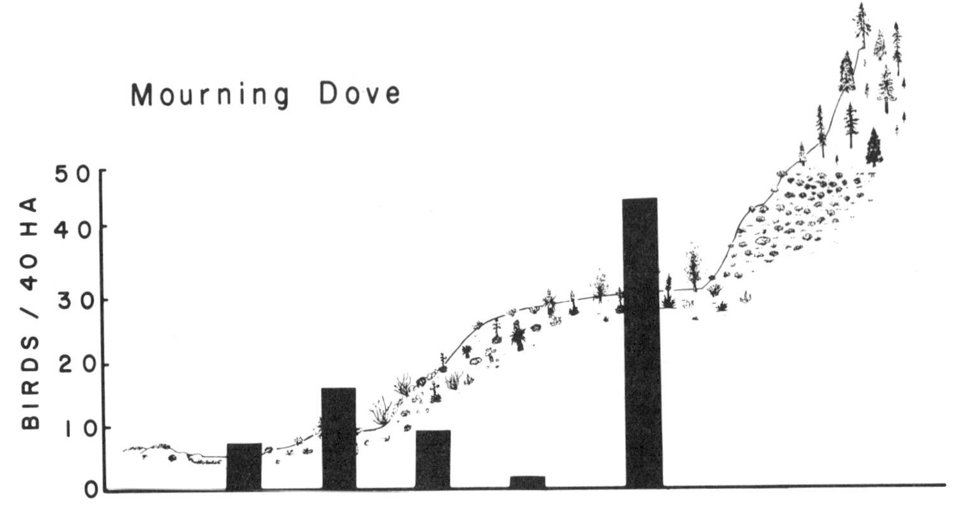

FIGURE 56 Mean annual Mourning Dove density at Deep Canyon. See fig. 38 for explanation of symbols.

FIGURE 57 Mourning Doves typically nest in trees, but occasionally a nest is built on the ground.

prevent overheating, and the adults must actively cool the eggs. To do this, the adult's brood patch is placed firmly against the eggs, and the absorbed heat is dissipated evaporatively by the adult's very effective gular flutter. Nests on the ground are exposed to even higher air temperatures than those in trees and may present special problems to the adults. Ground-nesting birds must use prodigious quantities of water for evaporative cooling during hot weather. But because Mourning Doves are strong fliers, they can replenish their water supplies frequently throughout the day.

Ground Dove. *Columbina passerina*
34–40 grams

Synonyms: Mexican Ground Dove; *Columbigallina passerina.*

Range: From southern California eastward to South Carolina, thence southward to Costa Rica. Also in South America from Colombia and Venezuela to Ecuador and Brazil.

Deep Canyon: Permanent resident throughout the valley floor and alluvial plain. Abundant in urban areas.

The species name, *passerina,* is Latin for "sparrowlike" and is a reference to the Ground Dove's diminutive size—only a little larger than a Golden-crowned Sparrow. Ground Doves are gregarious and occur in flocks, even during the breeding season. They are most common around parks and towns, usually nesting in close proximity to water (Bent 1932). They are decidedly uncommon in the pristine, dry desert regions of the Deep Canyon Transect.

Ground Doves are very aggressive, and their agonistic behavior is easily witnessed where they concentrate. Direct attacks are frequent, but more commonly a dominant individual merely chases other birds with head lowered and back hunched. Ground Doves use their rufous wing patches to signal threats and frequently flick their wings during combative encounters. Wing flicking also signals readiness to breed, and its occurrence throughout much of the year suggests a protracted nesting season. In Imperial Valley, to the south of Deep Canyon, Ground Dove nests have been found in December and February (Bent 1932). Breeding starts in winter and several broods (possibly four) are produced each year (Harrison 1978). Ground Doves may nest throughout the year at Deep Canyon. The nest is usually situated in a tree or shrub (occasionally on the ground) and is quite substantial for a dove.

CUCKOOS—CUCULIDAE

The 127 species in this family exhibit great superficial diversity. All of them, however, have two toes directed forward and two directed backward, a condition that they share with parrots and some woodpeckers.

The toe arrangement gives ground-dwelling cuckoos, like the Roadrunner, a unique X-like footprint. The bill of cuckoos is moderately long, slightly decurved, and usually rather stout. The tail is always long. All four North American species build their own nests and rear their own young, unlike certain parasitic species found elsewhere.

Roadrunner. *Geococcyx californianus*
300 grams

Synonyms: None since 1900.

Range: Resident from northern California, Nevada, southern Utah, Colorado, southwestern Kansas eastward to northwestern Louisiana, thence south to Baja California and as far south as Puebla and Veracruz, Mexico.

Deep Canyon: Resident from the valley floor up-mountain to the coniferous forest.

Roadrunners are quintessential desert birds possessed of a peculiar half-reptilian aura. They thrive without drinking water, even in the harshest of environments, and bask in the sun like lizards. Each pair mates for life and lives on its territory year-round, defending it throughout winter to insure an adequate food supply (Ohmart 1973). Their diet consists of some vegetable matter and whatever animals they can catch and swallow whole, including birds, lizards, arthropods, and rodents as large as wood rats (*Neotoma*) (Bent 1940).

A courting male's call is quite bizarre. It consists of a series of moaning *coos*, descending in pitch and slowing down toward the end, sounding much like a person in extreme pain or ecstasy. The call is given from a high perch with the bird's back hunched and its head pointing toward the ground. Each *coo* is pumped out with what seems to be a great effort. The call is loud, carries quite far, and is repeated monotonously.

Deep Canyon's Roadrunners seemingly nest in all habitats except the coniferous forest. Breeding probably begins in February or March in the Lower Sonoran Life Zone and about a month later in the Upper Sonoran Life Zone. On 4 April 1979, Roadrunner nests in the desert washes contained young, while on Pinyon Flat, males were still engaged in courtship calling.

Around Tucson, Arizona, the Roadrunner has a bimodal nesting pattern: mid-April to mid-June and late July to mid-September (Ohmart 1973). The breeding pause corresponds with the period of extreme heat and aridity. It is followed by increased nesting activity after the late July rains. Summer rains occur less predictably at Deep Canyon than in southern Arizona. Thus, in Deep Canyon's Lower Sonoran Life Zone, Roadrunners might produce only a single clutch in most years but may re-nest in years with significant summer rains.

The somewhat bulky stick-nest is usually situated 1 to 3 m above the ground in a low tree, thicket, or cactus clump (Bent 1940). All six of the

FIGURE 58 Typical Roadrunner nests at Deep Canyon are bulky stick-bowls placed in a dense palo verde tree. Sometimes Roadrunners nest on the ground.

nests I found at Deep Canyon were built in mistletoe clumps in palo verde trees (fig. 58). Roadrunners lay two to six eggs, apparently at a somewhat irregular interval (Ohmart 1973), and incubation begins with the first egg. Consequently, nestlings of different ages result. Asynchronous hatching serves to distribute the young's food demands over a broad period. If the food supply diminishes, the adults eat the younger chicks (Bent 1940). Initially, nestlings are fed insects, but within a few days the diet shifts almost exlusively to lizards. Frequently, such large lizards are poked into the young bird's gape that the nestling settles back to begin digestion with the lizard's hind legs and tail protruding from its mouth.

Desert adaptations: The Roadrunner is a relatively large, essentially flightless, diurnal carnivore endemic to a region of very low primary productivity. Other avian desert carnivores are either smaller, nocturnal, or strong fliers. The Roadrunner's niche is thus unique, and the bird has evolved several remarkable adaptations for coping with the multiple problems of heat, aridity, and poor food supply.

Problems of energy conservation are partly met by having clutch size regulated by food supply, reducing body temperature at night (hypothermia)—thus reducing the rate of metabolism—and using solar radiation (a source of thermal energy) by basking. In the morning, Roadrunners with low body temperatures typically position themselves in the open, with their backs toward the sun. They spread their scapular feathers, exposing bare, black skin. The black skin absorbs the sun's heat, and body temperature increases without metabolic cost. Roadrunners bask at other times, and studies using artificial solar radiation (Ohmart and Lasiewski 1971) show that by basking, they can maintain a basal metabolic rate at air temperatures as low as 9° C. Without solar radiation, metabolism begins to increase if air temperature falls below 27° C (Calder and Schmidt-Nielsen 1967). Roadrunners, therefore, reduce their food demands by adopting the heliothermic habits of their major prey base, the reptiles. Basking should be most important during the winter months when lizards and arthropods are less available and air temperatures are low.

Roadrunners prey on such passerines as Dark-eyed Juncos, House Finches, and House Sparrows. Zimmerman (1970) theorized that Roadrunners may subsist on birds during the winter months, if insects and lizards are unavailable. The winter influx of sparrows at Deep Canyon may well be important to the resident population of Roadrunners.

Adaptations related to improved water economy include: (1) the use of gular flutter to increase efficiently evaporative cooling during heat stress; (2) the elimination of salts via the nasal salt gland (Ohmart 1972), thereby avoiding the water loss that would result if salts were eliminated by the kidneys; and (3) the consumption of nestling fecal material by the adults. The nestling's urine and feces contain more water than does adult Roadrunner excreta. Consequently, the parent birds can satisfy 7.5 percent of their daily water requirements by consuming fecal sacs during nest sanitation (Calder 1968).

BARN OWLS—TYTONIDAE

Though closely related to the typical owls (Strigidae), Barn Owls differ slightly in skeletal structure and have a facial disk of pale, stiff feathers centered about the eyes. All nine members of the family are nocturnal hunters.

Barn Owl. *Tyto alba*
385–530 grams

Synonyms: *Strix pratincola; Aluco pratincola.*

Range: Cosmopolitan.

Deep Canyon: A few occur on Pinyon Flat but are more common at lower elevations around towns. Frequently roost in palm trees with well-developed skirts.

Barn Owls nest in cavities of all types, including crevices in rocks, outcrops, and cliffs. Because rodents reach their greatest density on Deep Canyon's rocky slopes (Ryan 1968), the adjacent gorges would seem ideal sites for owls. I have not found Barn Owls in these locations, however, and this may be because of the presence of Prairie Falcons—known to persecute owls—in the canyons.

The Barn Owls' acute sense of hearing enables them to locate and capture prey in total darkness. If prey is abundant, breeding may occur at any time of year. Barn Owl reproduction exhibits some remarkable adaptations for unpredictable food supplies. The clutch size is highly variable (three to eleven), with more eggs laid in good years. Eggs are laid at two-day or longer intervals, and incubation begins with the first egg. Hence, young of greatly differing ages result. Given abundant prey, most young survive to fledge (sixty days). If food is limited, only the oldest and largest chicks survive. This flexibility in reproduction contributes to the Barn Owl's remarkable success.

TYPICAL OWLS—STRIGIDAE

Six species of typical owls are known from the Deep Canyon Transect. Three of these are permanent residents, two are winter visitors, and one—the Elf Owl—is of accidental occurrence. Although not found during this study, two other species of strigids probably occur as sparse residents in the coniferous forest—the Saw-whet Owl (*Aegolius acadicus*) and the Pygmy Owl (*Glaucidium gnoma*).

Like all owls, strigids have dense, soft plumage. The feathers' ragged edges give the owls virtually soundless flight, enabling them to arrive suddenly and silently at night. This peculiar trait led Native Americans to associate owls with death. The association of owls with wisdom in European mythology has no similar biological basis. Much of our information about the diet of owls comes from examining their castings—regurgitated pellets of indigestible material such as fur, teeth, bones, and insect exoskeletons. (Many other species of birds produce castings, including shorebirds, flycatchers, hawks, and shrikes.) Most strigid owls are nocturnal. A few, like the Short-eared Owl, are regularly active during the day, particularly when the sky is overcast.

Screech Owl. *Otus asio*
Males 140 grams; females 155 grams

Synonyms: Little Red Owl; Mottled Owl; *Scops asio.*

Range: Common and widespread resident in North America from Alaska and southern Canada south to southern Mexico.

Deep Canyon: Permanent resident. Nests in coniferous forest and piñon-juniper habitats, and in palm trees on the alluvial plain and valley floor.

The Screech Owl is one of North America's most nocturnal owls. It nests in tree cavities, usually ones hollowed out by woodpeckers. Thus, at Deep Canyon, Screech Owls are most abundant in the coniferous forest and piñon-juniper habitats. Nesting begins in late February or early March, and males remain territorial for about ten months out of the year. They readily respond to an imitation of their call, a series of six to ten low, mellow hoots that start slowly and run together—something like a dropped Ping-Pong ball. Unlike the eastern race, western Screech Owls do not screech. The Screech Owl's diet consists mainly of small animals, although they are known to kill birds larger than themselves, including Rock Doves and Ruffed Grouse (*Bonasa umbellus*), and mammals as large as rats (Bent 1937).

Great Horned Owl. *Bubo virginianus*
Males 900 grams; females 1,150 grams

Synonyms: Pacific Horned Owl; Western Horned Owl.

Range: Throughout the Americas, except the West Indies, from the arctic treeline to the Strait of Magellan.

Deep Canyon: Permanent resident from piñon-juniper downslope to the valley floor.

Great Horned Owls find a variety of habitats suitable. Indeed, they are found in most localities that afford sheltered daytime roosts and nest sites, whether in trees or rock walls. The owl's large body size and water-rich diet enable it to live successfully in deserts year-round. Although individuals are occasionally flushed from their daytime roosts, the species is most often detected by its far-carrying hooted call. At Deep Canyon, Great Horned Owls are heard calling throughout the year. Males and females frequently call antiphonally. The female has the higher voice and generally more complex rhythm. The breeding season begins in late November or January, and eggs are laid in natural tree cavities, on rock ledges, or in rock or earth caves. Often the old nest of a Common Raven or Red-tailed Hawk is used.

Burrowing Owl. *Athene cunicularia*
Males 160 grams; females 145 grams

Synonyms: Western Burrowing Owl; *Speotyto cunicularia.*

Range: From southern Canada south to Tierra del Fuego.

Deep Canyon: From the floor of the Coachella Valley to the base of the Santa Rosa Mountains.

Grinnell and Swarth (1913) thought Burrowing Owls "very rare or entirely absent on the desert side of the [Peninsular] range." Development of the Coachella Valley may have increased their numbers, as they are now merely uncommon, rather than rare.

The Burrowing Owl's desert success results more from its behavior than from special physiological adaptations to desert conditions (Coulombe 1970, 1971). Especially important is its use of burrows for roosting and nesting. The burrow provides a more equitable environment than would an above-ground roost, offering relief from the desert's heat and aridity. It also, however, makes the owls more vulnerable to ground predators such as snakes, skunks, badgers (*Taxidea taxus*), coyotes, and feral house cats. Burrowing Owls sometimes excavate their own burrows but more often merely enlarge the preexisting burrow of a ground squirrel (*Citellus beecheyi*), desert tortoise (*Gopherus agassizi*), or badger. In optimum habitats, Burrowing Owls nest colonially with as many as ten to twelve pairs/ha. Information on the length of the breeding season is unavailable for Deep Canyon. In the nearby Imperial Valley, however, pair formation begins in mid-January, with egg-laying in late April to early May (Coulombe 1971). Some pairs apparently remain together throughout the year.

Coulombe (1971) described seasonal changes in the daily activity pattern of Burrowing Owls and related them to changing thermoregulatory and parental needs. During the year's cooler months, both members of the pair are frequently observed at the burrow entrance early in the morning and again late in the day. During midday, one owl remains at the burrow entrance while the other is below ground. In summer, the owl on midday sentry duty selects progressively higher perches as the air temperature rises and thereby avoids the hot ground-level air.

The Burrowing Owl's food habits have been reviewed by Coulombe (1970). He found that, although they take a wide variety of small vertebrates, they feed mainly on arthropods. On the Colorado Desert, earwigs (Dermaptera), crickets, tiger beetles, and various species of tenebrionid beetles comprise their principal prey. Foraging activity is usually observed at dawn and dusk. To what extent Burrowing Owls forage at night seems to be unknown. Foraging owls typically hover at heights of 5 to 10 m while searching for prey and devour captured prey on the ground before resuming the hover.

NIGHTJARS—CAPRIMULGIDAE

Also known as goatsuckers or nighthawks, caprimulgids are most closely related to the owls. Their small, weak bill, which contrasts markedly with their large head and eyes, opens wide to reveal a surprisingly large gape—an adaptation for catching flying insects. The gape's size is further enhanced by rows of rictal bristles along the upper mandible that deflect prey into the mouth. A special adaptation of the jaw muscles causes the mouth to snap shut, like a mousetrap, when prey hits the palate. Caprimulgids are most active at dawn and dusk and rest quietly during the day on the ground or along a tree branch.

Poor-will. *Phalaenoptilus nuttallii*
40–50 grams

Synonym: Desert Poor-will.

Range: Western North America from southern British Columbia eastward to southwestern Iowa and southward to central Mexico. Winters in southern part of range.

Deep Canyon: Permanent resident from chaparral down-mountain through the valley floor.

Although nocturnal birds were not censused during this study, Poor-wills appear to be most numerous in the chaparral. At dusk on 10 May 1979, I encountered six Poor-wills on a 3-km section of jeep trail that ascends through the chaparral on Santa Rosa Mountain's north face. There was a bright moon that evening, and stopping along the way, I heard many Poor-wills calling. The day before, it had snowed a few hundred meters higher on the mountain, but despite the cold, the Poor-wills were beginning to breed. Poor-wills on the warmer valley floor, in contrast, began breeding in mid-March.

Most Poor-wills forage at dawn and dusk on moths and beetles that they capture on the wing. Their acute night vision enables them to hunt in very dim light. Poor-wills, like all aerial insectivores, are dependent upon flying insects, which may become unavailable for periods of days or weeks during winter. Through evolution, aerial insectivores like swallows and flycatchers solved the problem of winter food shortages by migrating south. Poor-wills (Jaeger 1948) and some swifts adopted a different solution—hibernation. The natural circumstances that cause these birds to enter torpor are unclear but are probably associated with a negative energy balance. Because body temperature and metabolic rate are reduced during torpor, survival time on a fixed-energy reserve is increased. Bartholomew, Howell, and Cade (1957) measured a torpid Poor-will's metabolic rate and calculated that a 10-g fat reserve could sustain it for 100 days at 10° C. A nontorpid Poor-will, in contrast, would survive only ten days on an equal amount of fat.

Lesser Nighthawk. *Chordeiles acutipennis*
40–52 grams

Synonyms: Texas Nighthawk; Texas Trilling Nighthawk; Western Nighthawk.

Range: Breeds from the interior of central and southern California eastward through southwestern Texas and southward through Mexico to Chile and Peru. Winters southward from southern Baja California and Sinaloa, Mexico.

Deep Canyon: Migrant and summer visitor on the alluvial plain and valley floor.

The cryptically colored nighthawks blend in remarkably well with the desert pavement. During the day, they rest quietly on the open ground, even in the middle of summer. Thus, they are sometimes exposed to extreme heat. Like the Poor-will, Lesser Nighthawks can withstand the desert's heat partly because their efficient gular flutter cools them evaporatively with little increase in metabolic heat production.

Lesser Nighthawks forage on aerial insects and are sometimes attracted to the swarms of insects that gather around lights. As an adaptation to aerial feeding, their gape is very large, and the edge of the mouth is lined with hairlike bristle feathers that form a sort of insect sweep-net. Because they obtain water from their food, they do not require drinking water.

This species is most numerous at Deep Canyon during spring migration. During April and May, flocks of six to eight birds are often seen at dawn or dusk, flying low over the alluvial plain, golf courses, or residential areas. Their numbers decline in midsummer, but a few pairs (in most years, fewer than six) remain to nest on Deep Canyon's alluvial plain. Nesting birds choose flat, bare ground strewn with pebbles on which to lay their two cryptically colored eggs. No nest is built, and the female alone incubates the eggs, which hatch in eighteen to nineteen days. The semi-precocial young hatch down-covered and with their eyes open. Like the adults and eggs, they are cryptically colored, closely resembling pebbles. This helps conceal them from potential predators, such as coyotes, shrikes, and roadrunners.

Common Nighthawk. *Chordeiles minor*
69 grams

Synonyms: Pacific Nighthawk; Booming Nighthawk; Bull-bat.

Range: Breeds from the southern Yukon across northern Canada to Newfoundland south through Mexico to Puerto Rico. Winters in South America from Colombia and Venezuela south to central Argentina.

Deep Canyon: A single April record from the alluvial fan.

As is apparent from their range, Common Nighthawks are long-distance migrants, and like most birds of this type, they have long, pointed wings. In the field, they are distinguished from the Lesser Nighthawk by their larger size and by the white wing mark being positioned midway between the wrist and wing tip, rather than closer to the wing tip as in the Lesser Nighthawk. In California, the Common Nighthawk is decidedly a mountain species. It breeds in the Cascade-Sierra Nevada mountain area from Tulare County northward and in southern California only in the San Bernardino Mountains. The species is known to descend from the mountains to the valleys for evening foraging.

SWIFTS—APODIDAE

Apodidae is a Latinized form of Greek meaning "footless." It is formed from the prefix *a*, "without," and *pous*, "foot." The name is entirely appropriate, because swifts have remarkably small feet and very short legs. Normally they do not land on the ground or in vegetation, and they primarily use their feet for clinging to steep rock roosts. Swifts are the most aerial of all birds. Whereas other birds alight to bathe, drink, gather nest material, and copulate, swifts do all these things in the air. They sometimes spend days on end in rapid flight. Some species are even thought to sleep on the wing. Flight is energetically expensive for most birds, but swifts have evolved a number of adaptations that reduce their flight costs 50 to 75 percent (Hails 1979). These include a compact body; a streamlined shape; long, pointed, swept-back wings; and extensive use of gliding flight, which has a much lower cost than flapping flight. Swifts feed entirely on small animals—mainly insects—that they catch in the air. Reliable information on foraging is lacking, but they seem to locate concentrations of flying insects and then fly back and forth among them.

White-throated Swift. *Aeronautes saxatalis*
28–34 grams

Synonyms: White-bellied Swift; Rock Swift; *Aeronautes melanoleucus.*

Range: Southern British Columbia to Montana, northwestern South Dakota, Rocky Mountains, southwestern United States south to Guatemala and El Salvador. Winters from central California, central Arizona, and southwestern New Mexico southward.

Deep Canyon: Permanent resident, occurs throughout the transect.

White-throated Swifts (fig. 59) are usually detected by their twittering, chattering call. They vocalize at all times of the year and may use their calls to maintain contact while foraging. Like other swifts, this species feeds entirely on the wing. Certain meteorological conditions, such as vernal winds, seem to produce concentrations of "aerial plankton" at the mouths of desert valleys. These sometimes attract large numbers of swifts. One such concentration occurred on 10 February 1979 when I saw 170 White-throated Swifts flying back and forth at about 1,200 m elevation over Deep Canyon. Occasionally Violet-green Swallows and Vaux's Swifts join White-throated Swifts foraging over the canyon.

White-throated Swifts roost and nest in narrow rock crevices. Several swift colonies probably occur at Deep Canyon, because suitable roost sites are abundant. I found one colony in Carrizo Canyon near Black Hill on 30 April 1977. It was located in an inaccessible crevice above the first major falls one encounters walking up the canyon from Highway 74. Many swifts entered and left the colony while I watched from about 7 m away. They were not carrying nesting material but may have been feeding nestlings. Harrison (1978) reports that White-throated Swifts lay four to five eggs. No details exist concerning the incubation period, the nestlings, or nestling period.

White-throated Swifts are present at Deep Canyon throughout the year, but their activity in winter is limited to periods of warm weather. During cold spells, airborne insects are scarce, foraging becomes unprofitable, and the swifts retreat to their rock crevices to wait out the weather in a torporous state. Like caprimulgids and hummingbirds, swifts are capable of greatly reducing their energy demands by becoming torpid (Koskimies 1948, Bartholomew et al. 1957). In the laboratory, White-throated Swifts enter torpor more readily when they have lost weight, suggesting that

FIGURE 59 White-throated Swifts, common sights in the sky above Deep Canyon, nest in the canyon's rock crevices.

birds in the wild probably stop foraging when their energy balance becomes consistently negative. White-throated Swifts can crawl about at body temperatures as low as 22° C (72° F) and are fully active at body temperatures as low as 35° C, which is unusual for most birds.

White-throated Swifts frequently drink by skimming the water's surface. On a hot summer day, while I was immersed up to my neck in one of Deep Canyon's pools, several White-throated Swifts wheeled down from above to drink within 2 m of me. That they were coming to drink, and not to catch insects on or near the water's surface, was quite clear. Because most insectivorous birds obtain all the water they need from their prey, what might account for the swifts' habit of drinking? Perhaps swifts hold their beaks open while foraging. If so, the rapid air movement over the moist buccal membranes could cause considerable water loss, possibly more than can be replaced by dietary intake alone.

During the spring, White-throated Swifts engage in aerial displays and copulation. Display flights involve glides with wings held stiffly upward at a distinct angle. Copulating individuals cling together in a twittering, tumbling free fall and occasionally strike the ground before parting.

HUMMINGBIRDS—TROCHILIDAE

In their day-to-day living, the outstanding characteristics of hummingbirds are their curiosity, fearlessness, and pugnacity. They readily approach people, even in the wild, and not infrequently land on one's head or shoulder. They vigorously attack birds many times their size and show no mercy toward their own kind. Their color and antics capture the interest of all who encounter them and led Buffon to comment that nature had blessed "them with all the gifts of which she has only given other birds a share."

Hummingbirds are found only in the New World where 321 species, grouped into 121 genera, are known. Although they range from Tierra del Fuego to southern Alaska, their center of greatest abundance is the northern Andes. This suggests that the family evolved in South America (Mayr 1964*b*).

Hummingbirds range in size from the smallest known bird, the 2-g Bee Hummingbird (*Mellisuga helenae*) of the West Indies, to the 20-g Giant Hummingbird (*Patagona gigas*) of the Andes. As a consequence of their small size, hummingbirds lead intense lives. They must feed frequently throughout the day, visiting flowers from just before dawn to just after sunset.

Misconceptions concerning hummingbirds abound. Among the more common are that they have no feet, never land, have hollow tongues, eat only nectar, see only red, and mate in midair.

All North American hummingbirds have long, pointed bills and an extensible tongue. Legend held that the hummingbird's tongue was hollow, which allowed them to suck up nectar. Careful studies by Weymouth et al. (1964) revealed this to be false. Instead of being hollow, the solid tongue forks distally, forming two membranous troughs with fimbriated (fringed) edges. These narrow troughs enable the birds to take up nectar by capillary action.

In temperate latitudes, red is the commonest color of hummingbird-pollinated flowers (Grant and Grant 1968), and there persists a common belief that red preferentially attracts hummingbirds. Recent behavioral studies (Goldsmith and Goldsmith 1979) demonstrate that hummingbirds are not preferentially attracted to red because of a higher visual sensitivity to red or an innate "releaser." Rather, they learn to associate the color with a nectar reward. In fact, blue can be as effective an attractant as red, if it is associated with a food reward. As stressed by Stiles (1976), reds and oranges are convenient flags, but their association with nectar sources is both learned and elastic. What, then, of the correlation between red color and pollination by hummingbirds? Goldsmith and Goldsmith (1979) postulate that, in the initiation of a predominately red hummingbird-pollinated flora, the initiative has fallen to the plants. Red does not play a unique role in attracting hummingbirds, but rather it is the color least likely to attract competing bees.

Hummingbirds not only feed on flower nectar but capture small, flying insects as well. Insects constitute an important, perhaps essential, source of proteins for growing nestlings. Female Anna's Hummingbirds preferentially select insects over nectar when feeding young (Carpenter and Castronova 1980).

The breeding behavior of all North American hummingbirds is similar. The female builds the nest, incubates the eggs, and rears the young with no direct assistance from the male. There is little or no direct contact between the sexes beyond actual mating itself. Because a male may fertilize several females and a female may mate with several males, the mating system is termed *promiscuous*. The following account of courtship and mating is drawn from my own observations and the accounts of Stiles (1973).

At the beginning of the nesting season, males become more aggressive, defend larger areas, sing more, and engage in chases of intruders that become more frequent and longer. Increased titers of testosterone presumably underlie the male's escalating intolerance of other birds. Males on territory seem to develop "hair triggers," and any hummingbird, females included, that enters a male's space is immediately attacked. Since females are attacked with the same vigor as intruding males, how does mating ever occur? The answer lies in the female's response to the attacking male.

The typical hummingbird response to an attacking territorial male is immediate retreat. If the intruder is an equally aggressive male, a face-off, followed by an upwardly spiraling combat flight, may occur. Receptive females show a different response; they perch passively. Initially the male tries to dislodge a perched female with short, darting attacks. If the female remains perched, the male gives one or more dive displays in which he climbs above the female and "stoops" on her. Copulation follows, and then the male and female go their separate ways.

The sequence of nesting events of the Anna's Hummingbird (Stiles 1973) is representative of North American hummingbirds in general. At the time of mating, the female already has her nest partly built but may continue to add material to it well into incubation. Nest construction may take from three days to two weeks, but the average is one week. The clutch is invariably two (as in all hummingbirds), and incubation requires 14 to 19 days. The nestling period ranges from 18 to 23 days. The young remain dependent on the female for one to two weeks after fledging. A female typically raises two broods per year and re-nests within ten days following a nest failure. Thus, a successful nesting attempt requires 46 to 70 days.

Black-chinned Hummingbird. *Archilochus alexandri*
2.6–4.1 grams

Synonym: *Trochilus alexandri.*

Range: Breeds in western United States, chiefly west of Rocky Mountains, from British Columbia south to northern Mexico. Winters from southeastern California south to southern Mexico.

Deep Canyon: Migrant and sparse summer visitor.

Female Black-chinned Hummingbirds typically nest in deciduous trees along stream bottoms, especially in canyons, but also find irrigated orchards suitable. According to Grinnell and Miller (1944), males prefer drier sites, "as up canyon sides, often amid live oaks and chaparral plants, or on desert washes where mesquite and catclaw [sic] thickets occur." At Deep Canyon, this species has not been seen in the chaparral. It is most frequently seen on the alluvial plain and near the Living Desert Reserve, where it has nested. Black-chinned Hummingbirds leave Deep Canyon in fall, but there is one old December record for Palm Springs (Grinnell 1904). Future winter sightings are possible.

Costa's Hummingbird. *Calypte costae*
2.4–4.0 grams

Synonym: Costa Hummingbird.

Range: Breeds from central California, southern Nevada, and southwestern Utah south to Sonora and Sinaloa, Mexico. Winters over most of breeding range, from southern California southward.

Deep Canyon: Permanent resident. Numbers increase in winter.

Costa's Hummingbirds occur at Deep Canyon from the valley floor upslope to the lower edge of the pines. They are most abundant on the alluvial plain and rocky slopes (fig. 60). The availability of two important food plants, chuparosa and desert lavender, makes these regions attractive to them. Chuparosas, an especially sought-after nectar source, are most numerous in the dry, desert washes, which is where the hummingbirds concentrate.

Chuparosa produces flowers at any time of year following sufficient rains. In many years, autumn rains produce a profusion of scarlet, tubular blossoms by November. Costa's Hummingbirds respond to the chuparosa bloom and reach their highest density at Deep Canyon in winter (fig. 61). Their numbers gradually decline throughout spring. By May, few chuparosa blossoms are available, and most *C. costae* leave Deep Canyon for the chaparral of the Coast Range to the northwest. Stiles (1973) noted that the spring arrival of *C. costae* in California's Santa Monica Mountains coincided with a decline in their numbers on the desert. He thought their migration was fairly direct, because the birds did not appear in the higher mountains between the Colorado Desert and the Coast Range.

Compared with other North American hummingbirds, the diminutive Costa's is extremely xerophilous—some individuals live year-round on the Colorado Desert. Midsummer desert residence is surprising since, because of their small size, Costa's Hummingbirds should have one of the highest rates of water loss of any endothermic vertebrate. How much water Costa's Hummingbirds require during a hot summer day is uncertain. Lasiewski (1964) found that a 3.2-g Costa's resting in the dark at 23° C

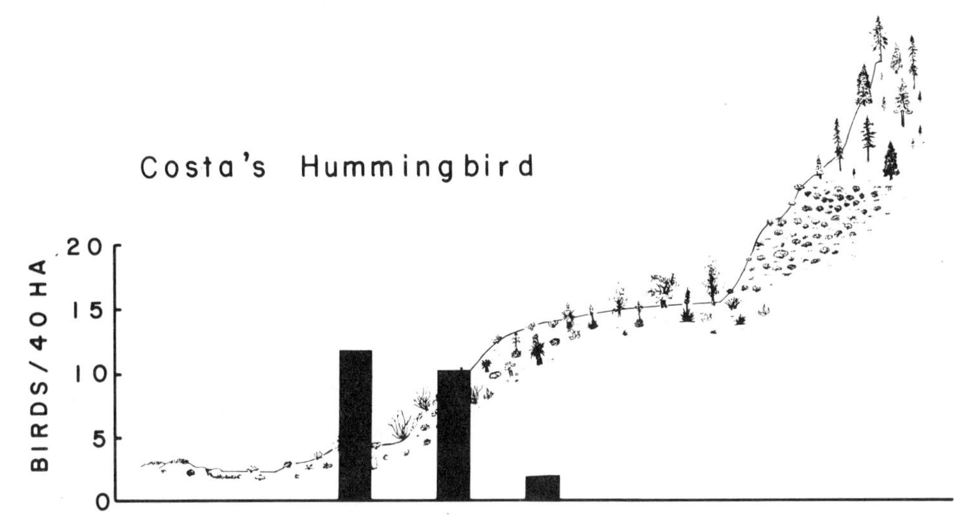

FIGURE 60 Mean annual density of Costa's Hummingbirds at Deep Canyon. See fig. 38 for explanation of symbols.

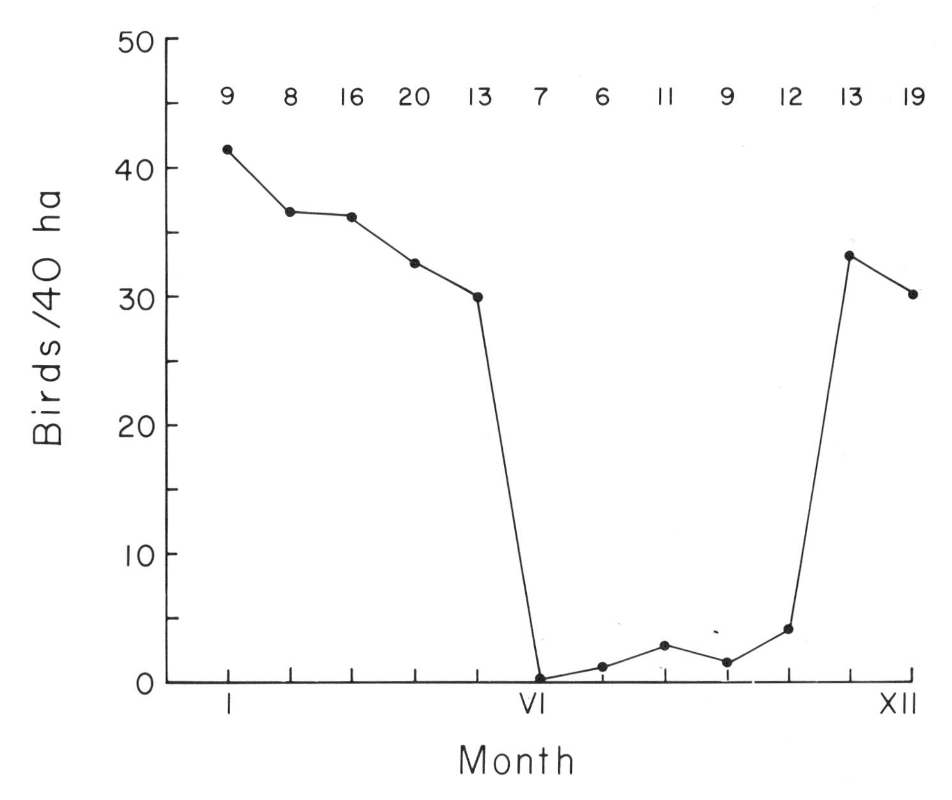

FIGURE 61 Density of Costa's Hummingbirds in Deep Canyon's dry desert washes. Values are means for the indicated number of strip transect censuses.

lost 0.93 g of water per day (29 percent of body mass/day). (A 70-kg [150-lb] man would need to drink 20 liters [5 gallons] of water per day if he lost water at the hummingbird's rate.) Since this represents the minimal rate, water loss on a hot summer day would be substantially greater. Whether Costa's Hummingbirds have any special physiological mechanisms for reducing water loss is uncertain. Because of the high water content of their nectar diet, such special mechanisms may not be needed.

On hot days, Costa's Hummingbirds appear to rely primarily on behavioral responses for thermoregulation. They become less active, retreat to the shade, and, to facilitate heat loss, compress their plumage and assume an upright cylindrical posture. Like other birds, they must resort to open-mouthed panting when air temperature exceeds body temperature. Accordingly, high temperatures are associated with high rates of water loss. Even though Costa's Hummingbirds apparently meet all their water needs with insects and nectar, they will drink and bathe if water is available. I have seen them drink from Deep Canyon Creek while hovering and, at other times, land at the water's edge.

Costa's Hummingbird is thought to be single brooded (Harrison 1978), yet this seems unlikely for the following reasons. Other species that lay two-egg clutches are usually either long lived (e.g., eagles and cranes) or produce multiple broods (e.g., pigeons and doves). In general, an inverse relation exists between longevity and body size—small birds lead short but intense lives. Thus, body size alone argues for more than one clutch per year. Furthermore, female *C. costae* re-nest quickly following a failure.

The belief that Costa's Hummingbird raises a single brood originated with Woods' (1927) statement that its stay on the Colorado Desert is insufficient to permit the raising of more than one clutch. At Deep Canyon, some Costa's Hummingbirds are present throughout the year. Thus, they have ample time to raise two broods. Most *C. costae,* however, leave the Deep Canyon Transect during the summer. Stiles (1973, p. 62) postulated that many of these migratory Costa's raise one brood in early spring in the desert and then move into the chaparral of the Coast Range in mid- to late-April to breed again. A similar pattern has been postulated for the Phainopepla (Walsberg 1977).

The nest of the Costa's Hummingbird is a small cup of feathers and a variety of plant material held together by elastic spider webs (fig. 62). This allows it to expand as the young grow. Nests are frequently placed near the ends of branches and are thus easily destroyed by wind. Partly because of this, Costa's nesting success (41 percent; Woods 1927) is lower than that of most open-nesting passerines (Lack 1954).

Anna's Hummingbird. *Calypte anna*
3.4–5.4 grams

Synonym: Anna Hummingbird.

Range: Breeds from northwestern California south to northwestern

FIGURE 62 Female Costa's Hummingbird on nest in smoke tree. The nest, bound together with spider webs, is elastic and expands as the young grow.

Baja California. Winters over the breeding range south on the Baja California mainland to Catavina and east across southern Arizona to northern Sonora.

Deep Canyon: Permanent resident. Numbers increase in winter as birds arrive from the north.

The breeding range of *C. anna* virtually coincides with the warmer and drier parts of the California chaparral (see Stiles 1973). During the breeding season (December to May), males establish territories in open terrain, while females nest in tracts of evergreen trees.

At Deep Canyon, as elsewhere in California, *C. anna* undergo a seasonal vertical migration. Following the breeding season, there is a midsummer movement up into the coniferous forest. In September and October, as autumn's cold cuts down the food supply in the pines, the birds return to the lowlands. At this time, there is a dispersal of the California population, and winter visitors begin arriving at Deep Canyon.

Costa's and Anna's Hummingbirds co-occur at Deep Canyon in the piñon-juniper and chaparral habitats. Males of both species will mate with females of either species, and hybrids are not uncommon (Wells et al. 1978). I have seen hybrids in the piñon-juniper that may represent local matings.

Calliope Hummingbird. *Stellula calliope*
2.6–3.4 grams

Synonym: *Selasphorus calliope.*

Range: Breeds in mountainous areas from central British Columbia and southwestern Alberta south to northern Baja California and east to Utah and western Colorado. Winters in Mexico south to Guerrero.

Deep Canyon: Sparse summer visitor of the coniferous forest.

This smallest of North American hummingbirds prefers high mountains for breeding. Birds begin arriving in California from their Mexican wintering areas in April, when the mountain slopes are still cold and snow covered. At this time, they are dependent on early blooming plants like manzanita (*Arctostaphylos* spp.) for nectar. Later they switch to indian paintbrush (*Castilleja* spp.), monkey-flower (*Mimulus* spp.), and penstemon (*Penstemon* spp.).

Few Calliope Hummingbirds have been observed on the Deep Canyon Transect. Only one nest, in the coniferous forest, has been found.

KINGFISHERS—ALCEDINIDAE

Worldwide, there are eighty-seven species of kingfishers. Both North American species are fishing kingfishers. That is, they dive into water after fish. Several southeast Asian kingfishers are terrestrial, however, and feed on lizards, snakes, and insects.

During the migratory season, Belted Kingfishers appear on the Deep Canyon Transect at golf course water hazards and sewage treatment plants.

WOODPECKERS—PICIDAE

Woodpeckers are specialized for extracting prey from wood and from in or under bark. They have straight, powerful bills that grow continuously to counteract their being worn down. They also have thick, stiff tail feathers, which help support the bird against vertical surfaces. The central pair of tail feathers, on which much of the support depends, are molted only after a new pair has grown in to replace them (Steinbacher 1964). In this way, the tail's function is preserved at all times. Highly adapted for a tree-climbing life style, woodpeckers' feet are equipped with strong claws and an outer front toe that is directed backward. This gives them two toes forward and two behind (an arrangement termed *zygodactylous* or *yoke-toed*).

A most remarkable feature of woodpeckers is their long, protrusible tongues. In some species, the tongue can extend 5 cm beyond the bill's tip. The tongues of species that feed on wood-boring larvae have sharp, spearlike, horny tips. Those of sapsuckers have brushlike tips for "lapping" up sap.

Most woodpeckers live in the same area year-round, but three species undertake annual migrations. Two of these, the Common Flicker and the Yellow-bellied Sapsucker, are North American species. The opportunistic Lewis' Woodpecker migrates irregularly to areas where nut-bearing trees are abundant.

Though woodpeckers mate for life, the pair bond dissolves following the breeding season. It must be reestablished each year, even in birds that winter on the same range. Drumming, loud rhythmic hammering of the bill on a resounding object, is an important prelude to pairing and functions like song in songbirds. After pairing, a nest site is chosen, with one sex generally doing the selecting. In the Yellow-bellied Sapsucker, Common Flicker, and Hairy Woodpecker, the male plays the principal role in nest site selection.

All North American woodpeckers nest in tree cavities that they excavate. Both sexes help excavate the nest, but the male does the larger share of the work. Some species (e.g., Hairy Woodpecker) make a new nest chamber each year, whereas others (e.g., Common Flicker) frequently reuse old nests. Many species excavate roost holes outside of the breeding season as well.

Raphael's (1980) in-depth study of cavity-nesting birds (including woodpeckers) in a western coniferous forest revealed that most nests are excavated in dead snags or in the dead portions of living trees. Most species preferred large, broken-topped snags with a good bark covering. The density of cavity-nesting birds was greatest at a snag density of eight trees greater than 38-cm diameter (breast height) per ha. Furthermore, Raphael found that the density of suitable snags limits cavity-nesting bird density, whereas snag-size diversity influences bird species diversity.

Woodpeckers, conspicuous and important components of forest ecosystems, are beneficial forest inhabitants. They do little damage to trees, create nesting cavities utilized by species that cannot drill wood (e.g., bluebirds, chickadees, and swallows), and can sometimes dramatically reduce populations of forest insect pests. Knight (1958) reported that woodpeckers reduced populations of Englemann Spruce Beetle (*Dendroctonus engelmanni*) by 45 to 98 percent. Because of their potential economic importance, woodpeckers are receiving increased attention. (For an introduction to the forest management aspects of woodpeckers, see Dickson et al. 1979 and Raphael 1980.)

Additional insights into woodpeckers are found in Lawrence's (1967) detailed natural history account of four species of woodpeckers: Hairy, Downy, Common Flicker, and Yellow-bellied Sapsucker.

Common Flicker. *Colaptes auratus*
130–160 grams

Synonyms: Red-shafted Flicker; *Colaptes cafer.*

Range: Southeastern Alaska and southwestern Canada south through Mexico.

Deep Canyon: Resident. Numbers augmented in winter by montane and (probably) out-of-state contingents of birds.

TABLE 41 Seasonal Variation in Common Flicker Density at Deep Canyon

Habitat	Birds/40 ha			
	Winter	Spring	Summer	Fall
Valley floor	3.76	0	0	0
Piñon-juniper	18.50	8.45	0	10.25
Chaparral	1.14	0	0	0.36
Coniferous forest	0	4.00	6.00	3.20

Flickers are permanent residents in Deep Canyon's piñon-juniper, chaparral, and coniferous forest habitats, yet there is considerable seasonal variation in their abundance. The census data (table 41) suggest that a down-mountain movement of flickers occurs in winter, coincident with the arrival of winter visitors from outside the transect. According to Miller (1951), this species highly prefers coniferous forest over piñon-juniper woodland. At Deep Canyon, however, the mean annual density of flickers is 2.6 times higher in the piñon-juniper than in the coniferous forest (8.93 versus 3.43 birds/40 ha).

Locally, Common Flicker nests have been found only in the coniferous forest, but a few pairs probably nest in the piñon-juniper habitat as well. Most of the nests I found were excavated between 2 and 30 m above the ground in dead jeffrey pines. Flickers usually lay 6 to 8 eggs (maximum 14). Although both sexes incubate during the day, only the male incubates at night. Flicker eggs hatch in 11 to 12 days, one of the shortest incubation periods of any North American bird. The young are fed by regurgitation, and the interval between feedings is relatively long (40 to 120 minutes). The young leave the nest 25 to 28 days after hatching and follow their parents for several weeks before attaining independence. Flickers are single-brooded but replace lost clutches.

Common Flickers drill fairly large nest cavities. Later, these provide roosts and nest sites for many birds that do not excavate their own holes, such as the American Kestrel and the Screech Owl.

Flickers have specialized for feeding on ants, which comprise about half their diet (Beal 1911). Their exceptionally long tongues are tipped with two small barbs and are coated with a sticky saliva to which the ants adhere. They forage extensively on the ground and frequently probe the earth for insects.

Acorn Woodpecker. *Melanerpes formicivorus*
60–80 grams

Synonyms: California Woodpecker; Mearns' Woodpecker; Ant-eating Woodpecker; *Balanosphyra formicivora.*

Range: Southwestern Oregon, California west of Sierra Nevada Mountains, Arizona and west-central Texas south to Panama.

Deep Canyon: Permanent resident in the coniferous forest.

FIGURE 63 Acorn Woodpeckers nest near the stands of golden-cup oak found in the Santa Rosa Mountains' coniferous forest.

Distinguished by its clownlike facial pattern (fig. 63), the Acorn Woodpecker is easily detected by its far-reaching call—*jack-a, jack-a, jack-a.* It is the most social of the North American woodpeckers and lives year-round in groups of two to fifteen birds of all ages and both sexes (for a detailed account, see MacRoberts and MacRoberts 1976). Members of the group cooperate in defending the territory, building the nest cavity, incubating the eggs, and feeding the young.

Throughout the summer, the Acorn Woodpecker eats mostly insects, often taken on the wing; but from fall to spring, it survives on acorns. It stores acorns in holes drilled in trees, utility poles, and wooden buildings. For this habit, the Spanish Californians called the bird *el carpintero.* The acorn larder is sometimes immense: one giant sycamore contained 20,000 acorns in its huge trunk (Dawson 1923).

There are at least two Acorn Woodpecker colonies in the transect's coniferous forest: one on Santa Rosa Mountain's north slope, near the old charcoal kiln on the "Whitman Road"; the other near the intersection of Garnet Queen Creek and the road to Santa Rosa Peak. Groves of golden-cup oaks (a favored food item) occur at both locations. Apparently scrub oaks constitute an inadequate food source, as the birds do not occur in the piñon-juniper habitat.

Yellow-bellied Sapsucker. *Sphyrapicus varius*
42–52 grams

Synonyms: Red-breasted Sapsucker; *Sphyrapicus ruber*; Sierra Yellow-bellied Sapsucker; Sierra Red-breasted Sapsucker.

Range: From southeastern Alaska, southern Canada, and Newfoundland south to northern Baja California and western Panama.

Deep Canyon: Scarce permanent resident and winter visitor.

The name *varius* is Latin for "varied" or "variegated" and refers to the plumage, which varies greatly within the species. The American Ornithologists' Union Nomenclature Committee currently recognizes five races of Yellow-bellied Sapsuckers, of which one—*S. varius daggetti*—occurs at Deep Canyon. Some authors (e.g., Short 1969, Mayr and Short 1970) maintain that this race is in fact a distinct species, the Red-breasted Sapsucker (*Sphyrapicus ruber*).

This sapsucker is encountered only occasionally in Deep Canyon's coniferous forest. A few pairs probably breed there, but as yet no nests have been found. Within its range, a considerable down-mountain movement of individuals takes place in winter. Accordingly, some sapsuckers seen at Deep Canyon between September and March may be winter visitors from other southern California mountains.

This species most often reveals its presence by the neat rows of holes that it drills in trees to tap the flow of sap. In addition to eating the sap that oozes from the holes, sapsuckers also eat the insects that the sap attracts. Downy Woodpeckers (*Picoides pubescens*), warblers, and hummingbirds are also said to find the sap attractive.

Hairy Woodpecker. *Picoides villosus*
60–70 grams

Synonyms: Cabanis Woodpecker; Cabanis HairyWoodpecker; *Dryobates villosus*; *Dendrocopos villosus*.

Range: Central Alaska to south-central Quebec and Newfoundland southward through mountains of Central America to western Panama.

Deep Canyon: Permanent resident of the coniferous forest and piñon-juniper habitats. Some coniferous forest birds drift down-mountain into the chaparral and piñon-juniper habitats in winter.

This woodpecker forages principally along the tree trunks, clear up to their tops, by chipping and hammering at the bark. Its stout bill allows the Hairy Woodpecker to extract the wood-boring larvae that make up the bulk of its diet. In addition to larvae, it eats ants, but not winged insects (Lawrence 1967).

Hairy Woodpeckers are most common in Deep Canyon's coniferous forest. Their numbers change seasonally with their combined fall and winter density averaging half their spring and summer density (2.27 versus 4.34 birds/40 ha). Two factors contribute to this change. First, although adults tend to remain on the same territorial range for life, they roam beyond their boundaries in fall and winter if food is scarce. Second, most juvenile Hairy Woodpeckers move away from the area of their birth with the approach of autumn.

Ladder-backed Woodpecker. *Picoides scalaris*
32–40 grams

Synonyms: Cactus Woodpecker; Texas Woodpecker; Saint Lucas Woodpecker; *Dryobates scalaris*; *Dendrocopos scalaris.*

Range: Southwestern United States south through Mexico to Chiapas, Mexico, and Belize.

Deep Canyon: Permanent resident below about 1,200 m.

The Ladder-backed Woodpecker is North America's smallest desert woodpecker. Small size allows it to nest in dead agave flower stalks, whereas the larger desert-dwelling woodpeckers, the Gila (*Centaurus uropygialis*) and Common Flicker, depend on giant saguaros for nest sites. Thus, the Ladder-backed Woodpecker can occupy desert regions from which the other species are excluded. In addition to agave flower stalks, it excavates nest cavities in tree-yuccas, palo verdes, cottonwoods, willows, mesquite, or, if none of these are available, telephone poles or fence posts.

Ladder-backs, nowhere abundant at Deep Canyon, were most numerous on the lower plateau (1.15 birds/40 ha). But even there, they were encountered during only 8.6 percent of the censuses. Grinnell and Swarth (1913) found them to be "conspicuously associated with the agave belt" and felt that they were restricted to the Lower Sonoran Life Zone. In contrast, I found several individuals in the piñon-juniper habitat on Pinyon Flat, where their range overlaps that of Nuttall's Woodpecker. None, however, have been seen on the valley floor of the Deep Canyon Transect, although they were found near Snow Creek and Cabazon by Grinnell and Swarth (1913).

Miller and Stebbins (1964) sometimes followed individual pairs for a mile or more through the Joshua trees. From this, they surmised that Ladder-backed Woodpeckers have large territories.

Nesting begins in April and continues through June. The typical clutch is four to five, but sometimes only two eggs are laid. Both parents incubate and help raise the young.

Nuttall's Woodpecker. *Picoides nuttallii*
35 grams

Synonyms: Nuttall Woodpecker; *Dryobates nuttallii*; *Dendrocopos nuttallii.*

Range: From northwestern California south to northwestern Baja California.

Deep Canyon: Permanent resident in the piñon-juniper habitat. A few nest in the oaks along Garnet Queen Creek at the lower edge of the coniferous forest.

Like the Ladder-backed Woodpecker, which it closely resembles, Nuttall's Woodpecker is decidedly uncommon at Deep Canyon. Individuals were encountered during 4 percent of the piñon-juniper censuses: their mean density was 0.4 birds/40 ha. Although they may nest in the piñon-juniper habitat, nests were found only near Garnet Queen Creek. Grinnell and Swarth (1913) also found evidence of nesting near Garnet Queen Creek.

Nuttall's Woodpecker is known to hybridize with the Ladder-backed Woodpecker (Short 1971) in several California localities. As yet, no hybrids have been found at Deep Canyon.

White-headed Woodpecker. *Picoides albolarvatus*
62 grams

Synonyms: Southern White-headed Woodpecker; *Xenopicus albolarvatus*; *Dryobates albolarvatus*; *Dendrocopos albolarvatus.*

Range: From south-central British Columbia south to southern California and western Nevada.

Deep Canyon: Permanent resident of the coniferous forest.

Easily recognized by its white head and large, white wing-patches, the White-headed Woodpecker is common in ponderosa and jeffrey pine forests of the western United States. At Deep Canyon, it was encountered on 28.6 pecent of the coniferous forest censuses and was the third most abundant woodpecker in that habitat. The species is somewhat nomadic in winter, when a few birds occasionally drift downslope into the chaparral.

This bird is so closely associated with pines that both its plumage and eggs are frequently soiled with pitch (Dawson 1923, Grinnell and Swarth 1913). Whether this adversely affects the young is unknown.

The summer diet consists mainly of wood ants (Grinnell and Swarth 1913, Dawson 1923), but spiders and other insects also are taken from beneath bark scales. The bird is a quiet forager, flaking and chipping off successive scale layers with angled strokes or using its bill as a lever rather than a hammer. The main trunk and proximal branches of pines are the White-headed Woodpecker's favored foraging sites.

In winter, this species forages extensively on pine seeds (Wetmore 1964) before the cones open (Tevis 1953). Tevis estimated that a group of White-headed Woodpeckers consumed 34 percent of the 1,656 cones on 20 sugar pines.

White-headed Woodpeckers are seen drinking more often than most woodpeckers (Grinnell et al. 1930, Ligon 1973), which Ligon suggests is associated with the high proportion of dry matter in their diet. Hence, they may require a water source within their range, but this is uncertain.

TYRANT FLYCATCHERS—TYRANNIDAE

The tyrant flycatchers originated in South America with a secondary radiation, like that of hummingbirds, into North America (Mayr 1964a). They are primitive songbirds related to the tropical American cotingas and manakins. Their resemblance to Old World flycatchers (Muscicapidae) results from convergent evolution rather than common ancestry.

Many species share the habit of catching flying insects out of the air. Perched quietly, they wait for an insect to fly past, dart out into the air after it, snap it up with an audible click of the bill, and then return to the same perch. Species foraging in this way have straight, broad bills, often with a hook at the tip, and rictal bristles at the mandible's base, which increase the gape's effective area. Bees and wasps are favorite prey.

Most tyrant flycatchers' nests are open, cup-shaped structures placed in a branch fork in a tree or shrub. The crested flycatchers (*Myiarchus* spp.) nest in cavities, however. Incubation is typically by the female alone, with the young fed by both parents.

Thirty-one species breed in the United States. Most migrate south to spend the winter in Central America or the West Indies. Thirteen species have occurred at Deep Canyon, and nine probably breed there.

Western Kingbird. *Tyrannus verticalis*
35–45 grams

Synonym: Arkansas Kingbird.

Range: Breeds from southwestern Canada south to Sonora, Mexico. Winters from the coast of South Carolina to Florida and in Middle America to northern Nicaragua.

Deep Canyon: Migrant and summer visitor.

Western Kingbirds are seen most frequently during the April migratory period, with no comparable increase during the fall migratory period (late August and early September). Kingbirds are scarce on the Deep Canyon Transect, with only a few pairs nesting locally.

Ash-throated Flycatcher. *Myiarchus cinerascens*
27–34 grams

Synonym: Northern Ash-throated Flycatcher.

Range: Western United States south from eastern Washington through Mexico, and east to Colorado and northern Texas.

Deep Canyon: Primarily a migrant and summer visitor, but a few are winter visitors.

Although this is Deep Canyon's most abundant tyrant flycatcher, it is never truly numerous. They are most common in late April and early

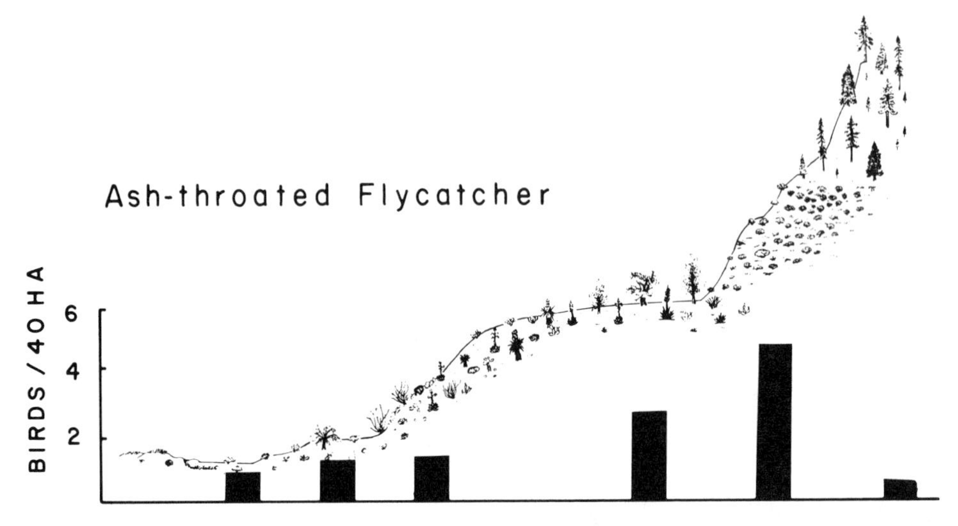

FIGURE 64 Mean density of Ash-throated Flycatchers at Deep Canyon during spring and summer (March through August). See fig. 38 for explanation of symbols.

May, when passing migrants swell the ranks of birds arriving to nest. During their period of summer residence, Ash-throated Flycatchers concentrate in the piñon-juniper woodland and chaparral (fig. 64). Even there, however, their density is low (3–5 birds/40 ha). Pairs are so isolated and widely spaced that they are seldom in sight or in hearing distance of each other. Their density is even lower at the lower elevations, where they must contend with the hottest summer weather. Seemingly, their major defense against heat is to become less active at midday. They do not drink water, relying instead on moisture obtained from their diet of bees, wasps, ants, robber flies (an enemy of honeybees), grasshoppers, and moths.

Ash-throated Flycatchers find a wide variety of cavities acceptable as nest sites. Holes in posts, pipes, and boxes are perfectly acceptable, but the more usual sites are natural cavities in trees and yuccas. Typically, four to five white-to-pinkish eggs are laid, and incubation is by the female alone. Nests have been found at Deep Canyon in the alluvial plain, lower plateau, piñon-juniper, and chaparral habitats.

Black Phoebe. *Sayornis nigricans*
16–21 grams

Synonym: California Black Phoebe.

Range: Western United States from California and Nevada south through Mexico.

Deep Canyon: Essentially vagrant but encountered nearly year-round.

Black Phoebes require mud for nest building and forage for insects over moist areas. Typical phoebe habitat consists of running water shaded by trees or canyon walls. The species is resident within its range but tends to wander about. It nests in the San Bernardino and San Jacinto Mountains (Grinnell and Miller 1944) but is probably prevented from nesting at Deep Canyon by the absence of mud. Typically the species is thinly scattered and, although encountered at Deep Canyon year-round, only a few individuals occur locally.

Say's Phoebe. *Sayornis saya*
18–24 grams

Synonyms: Say Phoebe; Rocky Mountain Say Phoebe; San Jose Say Phoebe; *Sayornis sayus.*

Range: Central Alaska and northwestern Canada south through Mexico. Generally resident, but during fall some birds leave the eastern and southern desert regions of California for the coast.

Deep Canyon: Permanent resident below 1,000 m.

Say's Phoebes are uncommon residents of Deep Canyon's lower elevations: average density is around one bird per 40 ha. Their ecological niche—open, sunny, arid terrain—is exactly opposite that of the Black Phoebe. For nesting, they only need crannies, in rock walls or buildings, that offer protection from the midday sun. The nest is made from weed stems, plant fibers, feathers, and hair (fig. 65). Unlike the Black Phoebe, Say's Phoebes do not use mud for nest building.

A color-banded female nested in the breezeway at the Boyd Research Center for over five years. The nest was always built in the same place, a narrow ledge beneath the roof (fig. 65). In 1978, this female laid her first egg on 26 March. In 1980, she laid her first egg on 29 March, followed by an egg each day until the clutch of five was complete. She began roosting on the nest at night after laying the third egg. On 18 April (18 days after the start of incubation), the nest contained two day-old nestlings, two pipped eggs, and one infertile egg. The typical incubation period of Say's Phoebes is 12 to 14 days (Harrison 1978). This female's longer period probably resulted from human disturbance, since she was sometimes flushed from the nest at night and did not return until morning. This undoubtedly resulted in cooling of the eggs and slowed their development. In 1980, the young fledged on 8 May, 20 days after hatching.

In most years, the female re-nested as soon as the young fledged, while the male continued feeding the fledglings. This pair's second brood was usually unsuccessful.

My notes on phoebe nestling behavior reveal something of their response to heat. Around 1400 (2:00 P.M.) on 5 May, air temperature near the nest (fig. 65) was 38.8° C (102° F). Panting, with mouths opened and wings drooped, all four young were perched outside the nest on the adjacent ledge. By 1630, the temperature had decreased to 35.5° C (96° F), and the birds were no longer panting. At 2030, the temperature

FIGURE 65 Nest and young of Say's Phoebe at the Boyd Research Center.

was 27.7° C (82° F) and the young were back in the nest huddled together. Air temperatures within the Boyd Research Center breezeway sometimes exceed 45° C (113° F). Although the young phoebes survive such heat, they must lose considerable amounts of water from panting. Whether they are more tolerant of dehydration than nondesert species is unknown.

Even though Say's Phoebes can stand bright sun and heat, they retreat to shade at midday. Their insect diet provides ample fluids and they do not drink, even if water is available. Like most flycatchers, phoebes regurgitate the indigestible portions of their prey as a pellet—a habit shared with many other birds, including owls.

Say's Phoebes forage closer to the ground than any of Deep Canyon's other flycatchers. They capture most insects from within half a meter of the ground, if not from the surface itself. Their lookout perch is typically a low bush or rock, from which they sally forth in buoyant flight. Their lightness of wing and preference for low perches may be adaptations to the strong desert winds (Grinnell and Miller 1944).

Dusky Flycatcher. *Empidonax oberholseri*
11 grams

Synonyms: Wright Flycatcher; *Empidonax wrightii.*

Range: Breeds from northwestern Canada south through western United States to southern California and Nevada, central Arizona, and northern New Mexico. Winters from southeastern Arizona south through Mexico.

Deep Canyon: Summer visitor in chaparral and coniferous forest habitats.

Empidonax flycatchers are characterized by their pale eye-rings and wing-bars, their small size, a nervous twitch of the tail (in all but the Gray Flycatcher, which dips its tail), and the *whit* call. Five species of *Empidonax* flycatchers occur regularly in California. Their field identification is difficult at best and sometimes impossible (e.g., distinguishing Hammond's from the Dusky Flycatcher). Dunn (1977) describes several ways of separating the western *Empidonax* flycatchers, of which the best is by their songs. Unfortunately, only the Western Flycatcher sings on migration. The others sing only on their breeding grounds.

In their survey of the Deep Canyon region, Grinnell and Swarth (1913) gave considerable attention to this species. They recorded it as breeding in abundance in the Santa Rosa Mountains down to the lowest edge of the Transition Zone, with "numbers of birds being seen and heard at the Garnet Queen Mine, altitude 6000 feet." They noted that although it was present on the higher slopes of Santa Rosa Mountain and Toro Peak, it was much less numerous there. The same situation still prevails today.

Gray Flycatcher. *Empidonax wrightii*
12 grams

Synonym: *Empidonax griseus.*

Range: Breeds from central Oregon south to southern California and east to central Colorado. Winters from southwestern United States south through Mexico.

Deep Canyon: Migrant and sparse summer visitor.

The spring migration of the Gray Flycatcher is inland, whereas the fall migration follows the coast. Spring migrants have occurred in Deep Canyon's piñon-juniper habitat only in small numbers. A few birds occur in the Santa Rosa Mountains' coniferous forest during the breeding season. One male captured near Stump Spring Meadow on 21 June 1979 had a moderate cloacal protuberance (indicating breeding condition) and seemed territorial. In recent decades, this species' breeding range has extended southward into the San Bernardino Mountains (Dunn 1977). Now it apparently extends into the Santa Rosa Mountains, although definite evidence of nesting at Deep Canyon is lacking.

Western Flycatcher. *Empidonax difficilis*
11 grams

Synonym: Northern Western Flycatcher.

Range: Breeds from southeastern Alaska south through the mountains to Baja California, and through the Rocky Mountains from Montana southward to Guatemala and Honduras. Winters from Baja California and northern Sonora south through Mexico to Central America.

Deep Canyon: Migrant and summer visitor.

FIGURE 66 Western Flycatchers are common spring migrants in the Colorado Desert's sandy washes.

As spring migrants, Western Flycatchers (fig. 66) are fairly common and widely distributed at Deep Canyon. They are regularly encountered in the dry desert washes, foraging under and about the foliage of palo verde trees.

This species' breeding status at Deep Canyon is unclear. Grinnell and Swarth (1913) found Western Flycatchers breeding in "small numbers" in the Santa Rosa Mountains and discovered a nest containing young on 26 June at Garnet Queen Mine. The nest was "placed in the end of a broken-off log, about eight feet above the bed of the stream." No nests were found during this study, and there are no June records. But the sighting of a Western Flycatcher along Garnet Queen Creek on 7 July 1979 by Steve Cardiff suggests a few pairs may still breed in the area.

Western Wood Pewee. *Contopus sordidulus*
11–13 grams

Synonyms: Wood Pewee; *Myiochanes richardsoni(i)*; *Contopus richardsonii*; *Contopus virens.*

Range: Breeds from central eastern Alaska east to southern MacKenzie, Canada, and south to Colombia. Winters from central Panama through northwestern South America.

Deep Canyon: Widespread as a migrant. Summer visitor in coniferous forest.

The Western Wood Pewee is the most common small flycatcher at Deep Canyon. During spring migration, which lasts from mid-April to early June, Wood Pewees occur on the Santa Rosa Mountains' desert slopes. They nest in the jeffrey pine forest and rank fifth in abundance of all summer visitors (see table 38). During the breeding season, the characteristic territorial call of the pewee, a harsh descending *pheer*, is heard throughout the day. The nest, built by the female, is a neat cup of fine grasses, bud scales, and bits of bark bound together with plant fibers, hairs, and, occasionally, spider webs. It is lined with fine grasses and feathers and is usually placed on a horizontal limb in a jeffrey pine. Incubation of the two to four white eggs is by the female alone, but the male helps feed the young.

Olive-sided Flycatcher. *Nuttallornis borealis*
27–33 grams

Synonym: *Contopus borealis.*

Range: Breeds from northern Alaska south to northern Baja California, across Canada, and in the mountains of eastern and western United States. Winters in northwestern South America.

Deep Canyon: Migrant and summer visitor.

Over much of their breeding range, Olive-sided Flycatchers are birds of the coniferous forest. During spring migration, small numbers pass through the low-lying regions of the Colorado Desert. At that time, individuals are found perched on low snags amid the cactus.

Typically, this bird fly-catches from the tops of tall trees, preferably dead ones that afford a better view. A few pairs nest at Deep Canyon near Garnet Queen Creek. The call is an imperative *tip three beers!*.

LARKS—ALAUDIDAE

Larks are ground-nesting birds of open or grassland habitats. The nest is placed in a shallow depression, usually in the shelter of a plant tuft or rock, with stones frequently added at the base to build up one side.

North America's only native lark, the Horned Lark, occurs infrequently at Deep Canyon. Most sightings probably represent winter visitors from more-northern breeding grounds. The species has bred nearby at Pleasant Valley and Joshua Tree National Monument (Miller and Stebbins 1964) and may eventually breed at the Ironwood Golf Course.

Horned Larks forage by walking instead of hopping. They feed on seeds during winter and insects in summer.

The Horned Lark's courtship flight is typical of larks generally. From his perch on the ground, the male soars skyward until nearly out of sight, sings his nuptial song while circling overhead, and then makes a grand plunge back to earth on folded wings, pulling out of the dive at the last second.

SWALLOWS—HIRUNDINIDAE

Swallows are widely regarded for their regular comings and goings. They are famous for returning to nest on the same day each year, such as at San Juan Capistrano, California. Alas, such punctuality is only a myth, as the return date varies by about two weeks. Such legends are outgrowths of the swallow's highly visible migrations, which take place during the day at low altitude.

Swallows, like swifts, are adapted for a life on the wing. They fly more than most birds, and their long, pointed wings make for economical flight. Indeed, flight costs swallows and swifts 25 to 50 percent less than other birds of similar size (Hails 1979). Swallows forage on the wing, coursing back and forth closer to the ground than swifts. They have short, flattened beaks and wide gapes suited for capturing insects in midair.

Seventy-five species of swallows occur throughout the world. Seven are known from California, and five occur regularly at Deep Canyon.

Violet-green Swallow. *Tachycineta thalassina*
15 grams

Synonym: Northern Violet-green Swallow.

Range: Breeds from central Alaska and southwestern Yukon south through Pacific coastal region to southern Baja California, and through Rocky Mountains and mountains of Mexico to Oaxaca. Winters mainly in Mexico and Central America, sparingly to southern Arizona and central coastal California.

Deep Canyon: Chiefly a migrant and summer visitor.

Violet-green Swallows, among the most beautiful and active of Deep Canyon's birds, forage higher than other swallows. They sometimes join White-throated Swifts over the canyon's inner gorge. Compared with swifts, they have throatier voices, make more abrupt turns, and can alight in trees or on wires.

Violet-green Swallows pass through the lower portions of the Deep Canyon Transect during their northern migration between mid-February to early May. Smaller numbers of southbound migrants make the return trip in October.

During winter, swallows seem to be nomadic and range widely in search of insect concentrations. Sometimes at Deep Canyon, weeks pass during winter without the sighting of a single swallow, only to have a small flock suddenly appear. There are no January records of swallows at Deep Canyon, but some individuals undoubtedly occur then.

The Violet-green Swallow, unlike many swallows, shows no special preference for water. Indeed, some birds nest in rock-walled desert canyons far from known water sources. At Deep Canyon, they nest in abundance in the Santa Rosa Mountains' coniferous forest, where they are the most numerous breeding species. They begin arriving in the coniferous forest in mid-April and soon seek out nest holes in trees. Violet-green Swallows find a wide variety of cavities acceptable as nest sites but prefer woodpecker holes. Where holes are numerous, loose colonies form, occasionally as many as twenty pairs in a single tree.

Tree Swallow. *Iridoprocne bicolor*
18 grams

Synonyms: White-bellied Swallow; *Tachycineta bicolor.*

Range: Breeds from north-central Alaska east across upper United States and Canada to Newfoundland; in west, south along the Pacific Coast to southern California. Winters from southern California and southern United States south to Honduras.

Deep Canyon: Migrant and possible summer visitor.

Climatically, this is the hardiest of North American swallows. In winter, it occurs as far north as New York.

Tree Swallows nest in woodpecker-excavated holes and prefer nest trees with bases surrounded by water or, at least, those that stand at the water's edge. Their aerial forage domain is above water or, at least, over ground that is damp, if not swampy. Few such situations exist within the Deep Canyon Transect, and it is doubtful that Tree Swallows nest there. Most individuals seen are apparently on migration.

Rough-winged Swallow. *Stelgidopteryx ruficollis*
15 grams

Synonyms: Northern Rough-winged Swallow; *Stelgidopteryx serripennis.*

Range: Southern Canada south to Argentina. Winters from southern United States southward.

Deep Canyon: Migrant and summer visitor.

The Rough-winged Swallow is less dependent on water than are other swallows and prefers localities that are extremely arid during the nesting season. It nests in burrows found in earth, sand, or gravel banks 1 to 15 m high, modifying holes dug by rodents. Pairs are solitary, and several pairs are rarely found in the same vicinity. A few pairs nest in the lower reaches of the Deep Canyon Transect; as, for example, near the Ironwood Golf Course.

Barn Swallow. *Hirundo rustica*
19 grams

Synonyms: Western Barn Swallow; *Hirundo erythrogastra.*

Range: North-central Alaska southeastward to Newfoundland and western Greenland, south to central Mexico. Winters from western Panama to southern South America.

Deep Canyon: Primarily a migrant.

Both Barn and Cliff Swallows build nests of mud, which limits their breeding distribution. Both species are found on the Deep Canyon Transect during the breeding season, but there is no evidence that they nest locally. They do not nest at the Boyd Research Center but may nest on buildings and freeway overpasses located near golf courses.

Cliff Swallow. *Petrochelidon pyrrhonota*
21 grams

Synonyms: Common Cliff Swallow; Greater Cliff Swallow; *Petrochelidon lunifrons*; *Petrochelidon albifrons.*

Range: Central Alaska southeastward to southern Quebec, south to central Mexico. Winters in central South America.

Deep Canyon: Migrant and possible summer visitor.

Cliff Swallows are early and abundant migrants through the Coachella Valley, with large flocks occurring in late February. The few individuals seen during summer might nest on developed portions of the Deep Canyon Transect or merely be late migrants.

CROWS, MAGPIES, AND JAYS—CORVIDAE

Corvids, the largest passerines, sometimes prey on other birds, taking eggs, young, and even adults. They are gregarious during the nonbreeding season. Most species are permanent residents; but some, like the Piñon Jay, are nomadic. They mate for life, build large stick-nests, and lack musical songs, their vocal repertory being limited to various calls.

Steller's Jay. *Cyanocitta stelleri*
110 grams

Synonyms: Blue-fronted Jay; Blue-fronted Steller Jay; Long-crested Jay.

Range: Southern Alaska and western Canada south through Mexico to southern Nicaragua.

Deep Canyon: Common resident of the coniferous forest.

Steller's Jays are common and conspicuous residents of the western Transition and Canadian Life Zones. Their normal voice, a raucous squawk, is lower than that of other jays. Like some other corvids, they are vocal mimics and give an excellent imitation of the *keeer-r-r* scream of the Red-tailed Hawk.

Grinnell and Swarth (1913) found Steller's Jays to be "exceedingly scarce" in the Santa Rosa Mountains. Today the jays are common there. Indeed, they are the fifth most abundant resident. The reason for this increase in jay numbers is uncertain.

In parts of their range, Steller's Jays descend from the mountains to the lowlands in winter and even venture out onto the desert (Phillips et al. 1964). Such movements are probably precipitated by heavy snowfall in the mountains. At Deep Canyon, where snowfall is moderate, Steller's Jays have not been recorded outside the coniferous forest.

Scrub Jay. *Aphelocoma coerulescens*
65–88 grams

Synonyms: Woodhouse's Jay; California Jay; *Aphelocoma californica.*

Range: Western United States south through Mexico; also in central Florida.

Deep Canyon: Permanent resident from the lower edge of the coniferous forest down-mountain to the upper portion of the lower plateau.

Although omnivorous like other corvids, Scrub Jays depend heavily on acorns for food. Each autumn they busily gather and store acorns against winter's famine, burying them at the base of a bush. Thus, they need not remember the exact location of each buried acorn but merely forage for them at the base of bushes. The Scrub Jay's distribution at Deep Canyon coincides with that of the scrub oak, reflecting their dependence on acorns.

Scrub Jays are formidable predators on other birds. They not only rob nests of eggs and nestlings but also capture and kill adult House Sparrows, House Finches, and White-crowned Sparrows, seizing them by one wing and beating them against the ground. At Deep Canyon, I have seen Scrub Jays rob oriole nests, and open-cup nesters generally lose many eggs to the ever-watchful jays.

Common Raven. *Corvus corax*
770–990 grams

Synonyms: American Raven; Western Raven; American Holarctic Raven.

Range: In North America, from the subarctic regions south through western United States, Mexico, and Central America to Nicaragua.

Deep Canyon: Sparse, wide-ranging resident.

Travel the Colorado Desert in summer, and you will eventually see a raven scavenging the highway amid a sea of shimmering heat waves. The vision of this black bird enduring heat intense enough to melt the road's tar immediately brings to mind the question: wouldn't a white coat be cooler? The answer is still a subject of debate, but recent evidence indicates that the raven's black coat may actually be cooler than a white one (see Walsberg et al. 1978). How is this possible, since black absorbs the sun's heat better than white? The explanation is that a black coat traps the heat at the surface rather than letting it penetrate to the skin. The surface feathers become very hot, but the heat is carried away by the wind. A white coat, in contrast, stays cooler at the surface but allows the sun's rays to penetrate deeper, bringing the heat close to the skin. This scheme is limited to large birds, as a thick feather layer is necessary to separate the hot surface from the skin. Appropriately, many large desert birds are dark—Turkey Vulture, Golden Eagle, and Common Raven—whereas small desert birds tend to be light.

Ravens are large, black birds with far-reaching voices. They typically travel in pairs and call to each other as they wing their way to distant appointments. These characteristics, combined with their habit of foraging along highways, make them conspicuous and create a false impression of abundance. This is true at Deep Canyon, where they are often seen along Highway 74 searching for road-kills.

Ravens are found throughout the Deep Canyon Transect but usually in low numbers, less than 5 birds/40 ha (fig. 67). Mean annual raven density in the piñon-juniper woodland during this study, however, was much higher than elsewhere (38.8 birds/40 ha). This high density resulted from an influx of ravens in late September and October. The rest of the year, raven density in the piñon-juniper habitat resembled that of other habitats (see chapter 9 for explanation).

Ravens typically nest on sheltered rock ledges or in large forks of trees. At Deep Canyon, they nest in the canyon's inner gorge, in the coniferous forest, and on large rock outcroppings. Birds roosting in the canyon move out onto the alluvial plain or valley floor during the day, forage, and then return to their roosts. Returning birds occasionally use "thermals" to ascend the mountain slope, circling lazily upward on outstretched wings in a manner similar to the Turkey Vulture's.

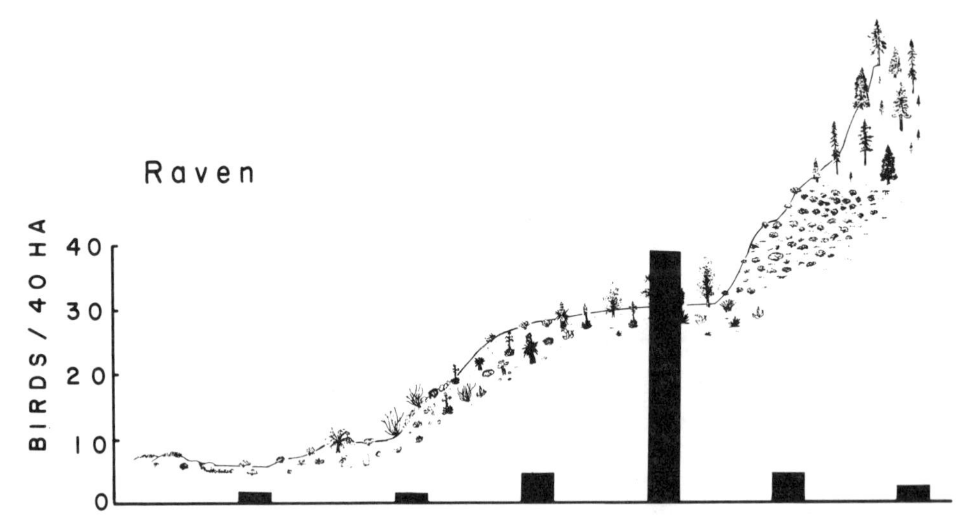

FIGURE 67 Mean annual density of Common Ravens at Deep Canyon. The high piñon-juniper value is the result of large flocks (some with over 125 ravens) congregating there in fall. See fig. 38 for explanation of symbols.

Piñon Jay. *Gymnorhinus cyanocephalus*
90–110 grams

Synonyms: *Cyanocephalus cyanocephalus.*

Range: Southwestern Saskatchewan and western United States south to Mexico.

Deep Canyon: Permanent resident of piñon-juniper woodland.

The Spanish name for this bird, *el piñonero,* suggests a strong attraction to the piñon tree. Throughout their extensive range in western North America, Piñon Jays inhabit primarily piñon-juniper woodland, but occasionally they are found in ponderosa pine forests as well. Although Piñon Jays occur year-round in Deep Canyon's piñon-juniper woodland, they are gregarious and nomadic. Many trips were made to the woodland without encountering a single jay, only to have a noisy flock of 100 or more jays suddenly appear on the next trip.

Reproductive cycles of most temperate-zone birds are keyed to seasonal changes in photoperiod. A few species use cues other than daylength and can breed throughout the year. The best known of these is the Red Crossbill, whose breeding is keyed to conifer seed production. It is said to have bred in every month (Tordoff and Dawson 1965). Piñon Jay reproduction is keyed to the piñon seed crop. The availability of piñon seeds permits the jays to breed in late winter (February) and again in late summer (August). In fact, breeding activities of Piñon Jays (from building nests to feeding fledglings) have been recorded in all months except December (see Ligon 1978).

Piñon Jays and piñon trees interact in a mutualistic fashion. The jays provide the trees with a primary means of seed dispersal, while at irregular intervals the trees provide the jays with abundant food. Piñon trees produce seed synchronously over large geographic areas. This has been interpreted by Ligon (1978) as an evolved mechanism that both overwhelms invertebrate seed and cone predators and increases the number of seeds per tree that will be cached by Piñon Jays and other vertebrates. Following a major seed crop, one large flock of Piñon Jays stores millions of seeds within its home range. Ligon (1978) found that seed-caching sites chosen by Piñon Jays are often conducive to germination and growth of piñon pine seedlings. Thus, areas cleared of piñon trees are "replanted" by the jays.

CHICKADEES, TITMICE, AND ALLIES—PARIDAE

Although "tits" resemble miniature crows, they probably are more closely related to nuthatches than corvids. The family, which consists of sixty-five species, occurs throughout the world (except Australia). In the New World, tits range from the arctic treeline south to the Guatemalan highlands. All are small (9 to 30 g), active, gregarious, and highly acrobatic birds of scrubland or woods. They are remarkably tame and intelligent. They have short, stout, pointed bills, which some species use for hammering seeds from pine cones. Parids are predominately insectivorous in summer, but depend on seeds to a considerable extent during winter. The plumage of male and female parids is similar in most species, which is characteristic of birds that mate for life. Most species lay large clutches.

Mountain Chickadee. *Parus gambeli*
11–12 grams

Synonyms: Bailey Mountain Chickadee; *Penthestes gambeli.*

Range: Mountains of the western United States from southwestern Canada south to Mexico.

Deep Canyon: Permanent resident of the coniferous forest.

The Mountain Chickadee's contact call, a clear *phoe´-be,* is heard year-round in Deep Canyon's coniferous forest, where this species is the second most abundant resident. In the piñon-juniper habitat, Mountain Chickadees are most abundant from mid-August until early April. Only a few scattered individuals occur there during summer. Most chickadees leave Deep Canyon's coniferous forest in winter, presumably for the piñon-juniper woodland. Those that remain behind forage with Pygmy Nuthatches in loose flocks. They reduce their nightly energy requirement by roosting communally in cavities.

The Mountain Chickadee's normal breeding habitat is pine forest of the Transition and Boreal Zones, but it occasionally breeds in piñon woodland. At Deep Canyon, it nests in abundance in the coniferous forest and possibly breeds in the piñon-juniper habitat as well. Although Mountain Chickadees sometimes nest under rocks or in holes in the ground, they usually prefer woodpecker holes or natural cavities in tree stumps. Like other parids, the Mountain Chickadee lays a large clutch (six to twelve), which hatches in fourteen days. The young are cared for by both parents.

Plain Titmouse. *Parus inornatus*
13–16 grams

Synonyms: Gray Titmouse; San Diego Titmouse; *Baeolophus inornatus.*

Range: Western United States (as far north as southern Oregon) south to Mexico.

Deep Canyon: Fairly common from piñon-juniper woodland upslope to the lower edge of the coniferous forest.

This plain gray bird with a crested head has a wide variety of calls. One is a chickadeelike *tsche-de-dee, tu-we-twee-twee.* Another, frequently heard in spring, is a slow, whistled *pee-wit, tee-wit, tee-wit.*

Plain Titmice are usually seen in pairs or singly, but family groups are sometimes encountered in summer. Unlike Mountain Chickadees,

Plain Titmice show no tendency to flock during winter. They forage among the smaller branches of scrub oaks, piñons, and junipers, using their stout bills to open piñon seeds or to chip away bark in search of insects. Tapping and pounding sounds accompany the birds as they move through the trees. Titmice occasionally forage on open ground, hopping about in search of insects.

Titmice nest in natural cavities or old woodpecker holes in piñons, junipers, and scrub oaks. The nest, made from a variety of materials, is built mainly or entirely by the female. Incubation is by the female alone. Bailey and Niedrach (1965) report that incubating birds sit very tightly, clinging to the nest lining if lifted from the nest.

Verdin. *Auriparus flaviceps*
6.5–7.5 grams

Synonym: California Verdin.

Range: Southwestern United States south through Mexico.

Deep Canyon: Resident from valley floor upslope to the piñon-juniper woodland.

Verdins are solitary little birds. They forage alone, sleep alone, and come together only during the breeding season. Together with hummingbirds and gnatcatchers, they are among the smallest desert birds. Because of their small size and resultant high energy demands, Verdins spend the majority of the daylight hours searching for food. Verdins mainly eat insects, but during the fall and winter months, they often feed on seed pods of palo verde and mesquite trees (Taylor 1971). They forage among the foliage of several plant species, using their feet skillfully to manipulate blossoms or hold large insect larvae.

During hot summer days, Verdins typically perch quietly in small patches of shade, a marked departure from their usually near-constant activity. Austin (1978) found that at low air temperatures (15° to 20° C) Verdins foraged 97 percent of the time. But above 35° C, foraging time decreased dramatically, falling to 22 percent between 40° to 45° C. The Verdins' decreased activity in the heat helps to conserve water.

Verdins produce a variety of noisy calls, quite loud in proportion to their size. The most common call is a slow *tee-chur, tee-chur-chur* repeated several times a minute at all seasons. It apparently functions as a contact note. When excited by a predator, Verdins give a churring *gee-gee-gee.* Their response to humans is a scolding *tschep*, repeated in a rapid staccato.

The Verdin is one of North America's most prolific nest builders. It constructs several nests each season—some weighing up to 109 g (Taylor 1971)—which are used both for breeding and roosting. Typical nests are spherical tangles of small sticks. They are lined with feathers (those of Gambel's Quail seemingly being preferred) and placed in the terminal branches of a tree or shrub (fig. 68). Roost nests are frequently smaller than breeding nests (Taylor 1971) and have shallower inner cups. Verdins

FIGURE 68 Verdin nests, built throughout the year, are used both for breeding and roosting.

enter their nests for the night earlier than most desert birds and are one of the latest birds to arise in the morning.

Nests offer some protection from inclement weather. For example, I found that eight out of ten Verdin nest interiors were dry after three days of winter rain (total 32 mm). Nests also offer some predator protection, which is not total, however, as 20 percent of the nests in Taylor's (1971) study suffered predation. Taylor never observed an act of nest predation but felt that snakes (especially coachwhips, *Masticophis*) were involved. Although snakes undoubtedly take some nestlings and eggs, several facts suggest that Cactus Wrens may also prey on Verdin nests. Cactus Wrens occasionally take vertebrate prey, including lizards and frogs (Bent 1948). Thus, they should have little trouble devouring a Verdin nestling. During the breeding season, both Verdins and Black-tailed Gnatcatchers scold and mob Cactus Wrens, a behavior typically directed toward predators. Furthermore, Cactus Wrens sometimes take over Verdin nests (Taylor 1971), which indicates that they can enter the narrow nest opening.

The Verdin's small size and exposure to the hot desert sun and wind would seem to make it particularly vulnerable to desiccation (see Bartholomew and Cade 1963). Nothing, however, is known of the Verdin's physiological responses to heat stress. Apparently Verdins can survive on the water they obtain from insects, as they have never been seen drinking.

Miller and Stebbins (1964) postulated that Verdins might retreat to their shaded nests during the heat of the day. I have never seen them do this.

Bushtit. *Psaltriparus minimus*
5 grams

Synonyms: Common Bushtit; California Bushtit; Lead-colored Bush-tit; *Psaltriparus plumbeus.*

Range: From extreme southwestern British Columbia south to Guatemala. From southern Oregon southeastward to westernmost Oklahoma and south to Sonora, Mexico.

Deep Canyon: Resident in piñon-juniper and chaparral habitats, vagrant elsewhere.

Bushtits, one of North America's smallest songbirds, are the size of Anna's Hummingbirds. They are active birds of rugged, dry, western canyons; and they frequent piñon-juniper, chaparral, and streamside habitats. They have long tails and drab, gray plumage. The sexes are alike, except that adult females have light, straw-colored eyes; those of males and immatures are dark.

Except during the breeding season, Bushtits travel in flocks that may include from a few to fifty individuals. Because a single Bushtit family may contain up to fifteen birds, flocks probably represent one or more family groups. When moving through the scrub, flock members keep in constant vocal contact by uttering a high-pitched *tsit-tsit-tsit.*

At Deep Canyon, Bushtits nest in the piñon-juniper woodland and the chaparral. The gourd-shaped nest, woven of spider webs, lichens, hair, and plant fibers, is often located in plain view. The normal clutch varies from six to eight white eggs, but sets of twelve to fourteen are not unusual. Both sexes incubate, but the male apparently lacks a brood patch (Miller and Stebbins 1964).

NUTHATCHES—SITTIDAE

This family contains twenty-five species of small, mainly arboreal birds with slender, sharp-pointed beaks. Nuthatches are primarily insectivorous. Still, during fall and winter, they eat many seeds and nuts. Most species store food in holes or crevices in trees. The Sittidae is one of the few bird families whose members are known to use tools. The Brown-headed Nuthatch (*Sitta pusilla*) of the southeastern United States sometimes uses twigs to extract grubs from beneath bark (Morse 1968).

Nuthatches are remarkable for their tree-climbing abilities. Using long toes and strong claws, they climb upward or downward on tree trunks—always head first. They can even walk upside down under limbs. Unlike creepers and woodpeckers, nuthatches do not use their tails for support while climbing. Thus, their tails are generally soft and short.

White-breasted Nuthatch. *Sitta carolinensis*
16–18 grams

Synonyms: Slender-billed Nuthatch; Rocky Mountain Nuthatch; Carolina Nuthatch.

Range: Southern Canada south through Mexico, but by-passes most of the Great Plains area of the United States.

Deep Canyon: Resident of the coniferous forest. A few wander downslope as far as the piñon-juniper in autumn.

White-breasted Nuthatches are the largest of the North American nuthatches (fig. 69). They have rather slender bills and hammer less than other nuthatches. Unlike their smaller relatives, White-breasted Nuthatches do not excavate their own nest hole. Instead, they rely on natural tree cavities or old woodpecker holes. White-breasted Nuthatches are usually found as pairs or solitary individuals. Yet, in winter, they sometimes occur in mixed-species flocks with other tree-foraging birds.

White-breasted Nuthatches are considerably less common in Deep Canyon's coniferous forest than the smaller Pygmy Nuthatches. Although they are found in the piñon-juniper woodland, a habitat in which they sometimes breed, all nesting records at Deep Canyon are from the jeffrey pine forest.

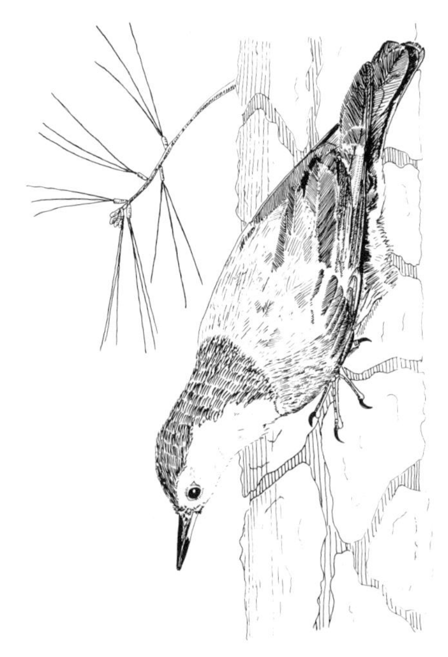

FIGURE 69 White-breasted Nuthatches walk up or down trees with abandon. They nest in the coniferous forest atop the Deep Canyon Transect.

Pygmy Nuthatch. *Sitta pygmaea*
9–11 grams

Synonyms: White-naped Nuthatch; White-naped Pygmy Nuthatch; *Sitta pygmea.*

Range: Mountainous areas from southern British Columbia south through western United States, east to southwestern South Dakota, and through Mexican highlands to Puebla.

Deep Canyon: Abundant resident of the coniferous forest.

This bird, encountered during every trip to the jeffrey pine forest, is one of Deep Canyon's most abundant residents. Its average annual density (90 birds/40 ha) exceeds that of most other residents. Throughout their extensive range, Pygmy Nuthatches are typically the most abundant winter resident in ponderosa and jeffrey pine forests.

Pygmy Nuthatches are noisy, busy, gregarious little birds, and this enhances one's impression of their abundance. They occur in pairs during the breeding season and in small flocks the rest of the year. They keep up a constant chatter as they fly from tree to tree, searching for food along tree trunks or the tips of pine branches. Skilled acrobats, they frequently hang upside down while exploring a pine needle cluster or ratchet head-first down a tree trunk while probing for insects in crevices and cracks in the bark.

Nesting begins in mid-April, and by early June flocks of young are everywhere. Pygmy Nuthatches nest in tree cavities that they excavate. The usual clutch size is five to nine, and the parents may be aided by helpers.

During fall and winter, Pygmy Nuthatches travel in flocks, frequently accompanied by Mountain Chickadees, and gather together in communal roosts (that may contain over 100 birds) in the late afternoon. Communal roosting helps protect the birds from the cold mountain nights. Knorr (1957) described one such early October assemblage in a ponderosa pine as follows:

> The tree in question was a very large yellow pine (*Pinus ponderosa*) which stood in the middle of a clearing. It had broken off about 30 feet from the ground, and the trunk, which remained standing, was pierced with holes. Subsequent investigation revealed it to be almost completely hollow, so that it consisted mainly of one very large cavity and a few additional minor cavities. As I sat at the edge of the clearing waiting for darkness, I heard the chattering of a flock of nuthatches approaching through the pines. It was apparent that the flock was quite large. Upon reaching the edge of the clearing, some individuals flew directly to the tree and entered the holes in the trunk. Others lingered in the surrounding pines investigating the bark and feeding. Eventually the remainder of the flock, individually and in small groups, flew to the old pine and disappeared into the holes. At first I tried to tally the incoming birds but lost count at 90. I finally estimated that there were at least 150 individuals rooting in the trunk, with a minimum of 100 sharing the same cavity. Whether this large number of birds represented a

single flock or a combination of several smaller flocks is unknown. The chattering continued for a short while and then all was quiet. Before leaving, I tapped the trunk and was rewarded with more chattering but no birds flew out.

CREEPERS—CERTHIIDAE

This family contains eight very similar species, named for their feeding behavior. They literally creep up the trunks and branches of trees. A foraging creeper slowly spirals up a tree trunk searching cracks for insects, insect eggs, and larvae. Then it flies to the bottom of another tree and starts the ascent again.

The creeper's long curved bill is specially designed for feeding on insects. During winter, many invertebrates seek refuge in the cracks and crevices of bark. They are safe there from stubby-billed gleaners like chickadees but are easily reached by the long-billed creepers.

Creepers of the genus *Certhia* share with woodpeckers strong, stiff tail feathers that support their climbing. As in woodpeckers, the central tail feathers are the most important, and they are molted only after the other, outer tail feathers have been replaced.

Brown Creeper. *Certhia familiaris*
8 grams

Synonyms: Sierra Creeper; Rocky Mountain Creeper; Mexican Creeper.

Range: Circumpolar in distribution. In North America, breeds from southeastern Alaska and southern Canada south through Mexico to Nicaragua.

Deep Canyon: Sparse resident of the coniferous forest.

The Brown Creeper has been aptly likened to an animated bit of bark. It is streaked brown and white, has a slender, decurved bill, and blends in remarkably well with its background (fig. 70).

Brown Creepers are decidedly uncommon at Deep Canyon, where they are apparently restricted to the coniferous forest. A day in the field produces at best only one or two solitary individuals. Grinnell and Swarth (1913) also found creepers to be quite rare in the Santa Rosa Mountains, encountering "probably not more than five or six altogether."

WRENTITS—CHAMAEIDAE

The Chamaeidae is North America's one unique family. It contains a single species, the Wrentit, which is found only in the west. Some authorities (Van Tyne and Berger 1976) include the Wrentit with the Old World Babblers.

FIGURE 70 When motionless, Brown Creepers resemble bits of bark.

Wrentit. *Chamaea fasciata*
16 grams

Synonym: Pallid Wren-tit.

Range: Western Oregon south to northern Baja California.

Deep Canyon: Fairly common resident of the chaparral.

Wrentits occur throughout the Santa Rosa Mountains' chaparral and occasionally wander a short distance into the adjoining coniferous forest. They also occur in the piñon-juniper habitat but are less numerous there than in the chaparral.

Among the most sedentary of birds, Wrentits live out their lives on a small home range (about 1 ha), where they hide deep within the brush. They hop through the brush like wrens and make short, weak flights from bush to bush. They dislike clearings so intensely that some birds refuse to cross firebreaks bulldozed through their territory.

Wrentits mate for life and are among the most "married" of birds. Pairs forage together, preen each other's feathers, and roost together at night with feathers interlaced and inner legs drawn up, appearing as one ball of feathers (Erickson 1948). Males and females call to each other throughout the year to maintain contact as they move through the dense brush. The male's song is a series of whistles that run together and end in a trill—like the sound made by a Ping-Pong ball dropped on a table. The female's song is similar but lacks the trill.

DIPPERS—CINCLIDAE

The semiaquatic dippers obtain their food largely under water, swimming and diving for insects, crustaceans, and small fishes in fast-running streams. They lack webbed feet but have dense, water-proof plumage underlaid with oily down. The Dippers that occasionally appear on Deep Canyon Creek probably arrive there from streams in the San Jacinto Mountains, where they breed.

WRENS—TROGLODYTIDAE

The family name is formed from the Greek *troglodytes*, meaning "cave dweller" and is an allusion to the wrens constant seeking of cover. Wrens are small, brown, active, and shy little birds. When disturbed, they sound their alarm with loud, harsh, and insistent notes. Their songs are surprisingly loud and marked by brilliance and sweetness. Most species sing throughout the year. Wrens spend much time near the ground, and most stay in dense cover. As an adaptation to life in the brush, they have short, rounded wings that give them excellent maneuverability. The sexes have similar plumage.

House Wren. *Troglodytes aedon*
8–9 grams

Synonyms: Western House Wren; Parkman Wren; Pacific House Wren.

Range: Breeds from southern Canada south to northern Baja California. In winter, south to Guerrero, Oaxaca, the Gulf coast, and southern Florida.

Deep Canyon: Common summer visitor in chaparral and coniferous forest habitats.

House Wrens arrive at Deep Canyon in March and depart in early November. Although these birds are generally restricted to the transect's higher elevations, individuals are occasionally seen on the alluvial plain. House Wrens seek out tree cavities, either natural or woodpecker-made, in which to build their compact cup-nests. At Deep Canyon, they prefer oaks for nesting and seem to avoid conifers. Two clutches of six to eight white or pink eggs are typical. Both sexes feed the young.

Bewick's Wren. *Troglodytes bewickii*
8–10 grams

Synonyms: San Diego Wren; Southwest Bewick Wren; *Thryomanes bewicki.*

Range: Southwestern British Columbia, central Washington southeastward to central Pennsylvania and south through Mexico.

Deep Canyon: Resident from the desert base of the mountains upslope to the coniferous forest.

FIGURE 71 At Deep Canyon, Bewick's Wrens are most numerous in the chaparral and piñon-juniper habitats.

Bewick's Wrens, shy, little birds that stay hidden from view in dense brush, are usually encountered singly or in pairs. They are easily distinguished from the similar House Wren by the white line above their eyes (fig. 71). Bewick's Wrens wag their long tails expressively, in a manner reminiscent of the African waxbills. They have a variety of loud, musical songs and sing throughout the year.

At Deep Canyon, Bewick's Wrens are most numerous in the chaparral (mean annual density 5.33 birds/40 ha) and nest from the lower limit of the coniferous forest downslope to the mountains' base. Bewick's Wren nests at Deep Canyon are typically placed in natural, scrub oak cavities within 0.5 m of the ground. One nest in the rocky slopes habitat, below the lower limit of scrub oaks, was placed in a rock crevice, however.

Bewick's Wrens are essentially resident at Deep Canyon, but some down-mountain drift occurs in fall and winter. A few wrens arrive on the alluvial plain in mid-August, spend the winter, and depart in mid-April. In good years, some of these "winter visitors" may stay to nest at the lower elevations, although Bewick's Wren nests have yet to be found on the alluvial plain. A male established a territory in Coyote Wash in 1980 and was still singing vigorously on 19 April. But by 7 May, he had departed, apparently unable to attract a mate. Nesting begins in March with the laying of the first of two clutches of five to seven eggs. Both sexes care for the young that, by early May, are following their parents through the brush.

Cactus Wren. *Campylorhynchus brunneicapillus*
38–42 grams

Synonyms: Northern Cactus Wren; *Heleodytes brunneicapillus.*

Range: Southwestern United States south to southern Mexico.

Deep Canyon: Resident from the valley floor up the Santa Rosa Mountains through the piñon-juniper habitat.

Like other members of the wren family, Cactus Wrens tend to be secretive and remain close to the ground. They seldom fly more than 50 m at a time and most often make shorter flights from bush to bush. Although a little time spent in the right habitat will reward one with a view of the wrens, it is common to first become aware of their presence by hearing their songs or calls or by seeing their conspicuous nests. The song, which carries a considerable distance, is a low, harsh *char-char-char-char* . . . , often repeated six to ten times in a mechanical monotone.

Throughout Deep Canyon, as elsewhere in their range, Cactus Wrens are closely associated with various species of cholla cactus. The wren's nest (fig. 72) is typically placed 0.5 to 1.0 m above the ground in a spiny cholla, which provides protection from predators. The large, gourd-shaped nests are made of grasses and lined with downy plant fibers and feathers. Each pair of wrens builds several nests per year, a trait shared with other members of the wren family. Nests, used both for roosting and breeding, are occupied year-round. In Arizona, Anderson and Anderson (1973) found the highest number of usable nests per territory in August and September, when surviving juvenile wrens began building their own roost nests. Between 1963 and 1967, the Andersons found an average of 4.7 nests per territory, with a lower number (3.6) between December and March.

Why Cactus Wrens are such prodigious nest builders is far from certain. They do not construct multiple nests as "decoys," as has sometimes been suggested, as all completed nests are used at some time by the male, female, or their young (Anderson and Anderson 1973). Although the nest produces a favorable microclimate, it seems unlikely that Cactus Wrens require protection from the weather: many more delicate species survive without the benefits of such elaborate homes. Nests in cholla are well protected from predators (especially owls), but other birds achieve similar protection by roosting in dense bushes.

Ricklefs and Hainsworth (1969) suggested that Cactus Wrens orient the nest opening to avoid the cool spring winds, while late-summer nests are oriented to take advantage of the breeze. Anderson and Anderson (1973) disputed the advantage of such an arrangement, however, and found no evidence of orientation when a much larger number of nests were examined (512 versus 63). Nest orientation at Deep Canyon appears to be constrained more by the shape of the cactus than by the climate.

Cactus Wrens occur in family groups from late spring through winter. But as the breeding season approaches, the adult pair drives off the young and retains sole possession of the territory.

FIGURE 72 A typical Cactus Wren nest in a jumping cholla.

In March, females at Deep Canyon are incubating their first clutch, while males occupy a secondary nest near the breeding nest. After the young reach independence, the parents may begin a second clutch in the male's secondary nest. Thus, one function of secondary nests may be to reduce the time-lag between clutches. Typically, four to five eggs are laid, and the female begins incubation before the clutch is completed, resulting in an asynchronous hatch (Anderson and Anderson 1973). The female alone incubates, but both parents feed the young.

Although Cactus Wrens have a high reproductive potential, producing up to fifteen young per year (three broods of five), mortality is high. Forty-one out of fifty-five banded nestlings disappeared within forty-five days after fledging (Anderson and Anderson 1973). Mortality is lower among adults, but few wrens live longer than five years in the wild.

During winter, wrens sometimes join flocks of White-crowned and Black-throated Sparrows, foraging with them on the ground in fairly open situations. The wrens forage by turning over bits of wood or other debris in search of insects. I saw one wren flip over a cow "chip" that weighed 94 g, more than twice the wren's weight. Such large pieces of debris must occasionally conceal insect concentrations to compensate the effort needed to overturn them.

The Cactus Wren's diet varies seasonally. During spring and summer, insects predominate, but some plant material (e.g., cactus fruits) is also eaten. In fall and winter, the proportion of plant material increases, and seeds become a major component of the diet. Consequently, Cactus Wrens obtain much less dietary water during fall and winter than in summer, and thus, they drink more frequently in winter (Anderson and Anderson 1963).

Cactus Wrens obtain relief from the midday summer sun by seeking shade. Ricklefs and Hainsworth (1968) found that in summer, wrens foraged in exposed situations only during the cooler morning hours. As air temperature rose, they foraged only in shady areas. During midday, foraging ceased altogether, and the wrens rested quietly in the shade with open beaks and drooping wings. At the Boyd Research Center, Cactus Wrens sought out the coolest microclimates in which to wait out the midday sun. Each noon found several wrens resting in the shade of the research center's breezeway. Four or five wrens congregated around the roof vent intercepting the cool air from the swamp cooler. By avoiding sunny situations when air temperatures are high, Cactus Wrens reduce their requirement for evaporative cooling and thus can survive on the water obtained from their insect prey.

How can Cactus Wrens land with impunity on cholla cactus that so easily impales unwary humans? Are their feet specially equipped to prevent penetration by the spines? Do they somehow avoid the spine points when landing? To answer these questions, I observed, at close range from a blind, wrens landing on cholla, and I also examined their feet. The sole of the Cactus Wren's foot is heavily calloused, but spaces between the callouses offer entry points for spines. Observing wrens land and take off from jumping cholla revealed that the spines do penetrate their feet—a distinct pop could be heard on takeoff as the spines pulled loose. When a spine penetrated deeply, the wren held the spine with its bill and lifted its foot free.

Sometimes Cactus Wrens become fatally impaled on cholla. Nestlings are especially vulnerable to impalement when they leave the nest and wander out onto the cactus branches. Many species of desert birds become impaled on cactus (see Miller 1936), and it is amazing that more do not meet death in this way.

Cañon Wren. *Catherpes mexicanus*
9–12 grams

Synonyms: Dotted Cañon Wren; Dotted Canyon Wren.

Range: Central coastal California, central southern British Columbia, and northern west-central United States south to southern Mexico.

Deep Canyon: Uncommon resident from the mountains' desert base upslope to the coniferous forest.

The cliff faces and boulder fields of steep-walled, steam-carrying canyons are the Cañon Wren's home. These entirely insectivorous birds forage for spiders and insects over shaded rock-faces and deep crevices. They nest in rock pockets. Never very numerous, Cañon Wrens are more frequently detected by the male's far-carrying song (a series of eight to twelve clear whistles of descending pitch) than by sight. They sing throughout the year, most often during spring. In summer, when the midday sun heats the canyon walls, the wrens retreat to shaded nooks. Although they frequently occur near water, some birds are found in dry canyons, suggesting that drinking may not be essential.

Rock Wren. *Salpinctes obsoletus*
14–17 grams

Synonym: Northern Rock Wren.

Range: South-central Canada south to Baja California and through the Mexican and Central American highlands to northwestern Costa Rica.

Deep Canyon: Widespread resident.

The generic name, *Salpinctes,* is derived from the Greek word for trumpeter and is an allusion to this wren's clear, vibrant song, a remarkable variety of trills. The most frequently heard call, especially in winter, is a loud, clear *ti-keer.*

Rock Wrens prefer open, rocky habitats that are more windswept and sunny than the shaded cliffs required by the Cañon Wren. They nest in a wide range of habitats, from dry, rock-strewn desert slopes to cool mountain forests. The one essential requirement for their presence is mammal burrows or rock crevices in which to build their nests.

At Deep Canyon, Rock Wrens nest from the mountains' base upslope to, or possibly through, the coniferous forest. They occur only accidentally on the valley floor and are found on the alluvial plain only between mid-August and early April. Winter visitors found on the alluvial plain could be either migrants from the northern portions of the species' range or birds from the higher regions of the Deep Canyon Transect.

The Rock Wren is the most subterranean of any of California's birds (Grinnell and Miller 1944). Its flattened body enables it to creep far into fissures. Indeed, birds have been seen to disappear down a mammal burrow and reappear at another opening several meters away. It seems likely that they retreat to burrows to escape the desert's midday heat.

One remarkable trait of the Rock Wren is its habit of paving the nest-burrow entrance with flattened stones. The function of the stone mosaic, which may cover an area two meters square and contain over 100 rock fragments, is unknown.

MOCKINGBIRDS AND THRASHERS—MIMIDAE

This Western Hemisphere family of thirty species is best represented in Central America, where about half the species are endemic. Many species have somber gray, brown, or blackish plumage, long tails, and rather short, rounded wings. Many Mimidae have strong, fairly long, decurved bills. Most feed on fruits, seeds, and various invertebrates such as insects, spiders, and centipedes. North American species inhabit dense scrubs. Some are excellent songsters and mimics of other species.

Mockingbird. *Mimus polyglottos*
48 grams

Synonym: Western Mockingbird.

Range: Southern Canada south to southern Baja California and the West Indies.

Deep Canyon: Resident below 300-m elevation, vagrant above.

At Deep Canyon, Mockingbirds (fig. 73) are most numerous near the orchards and gardens of the developed regions. A few occur year-round on the valley floor in association with mesquite thickets or higher up on the alluvial plain near water. At the Boyd Research Center, they are found in the dry desert washes from late September through late May. They apparently move farther down the alluvial plain toward water during the hot, dry summer. Only rarely are Mockingbirds encountered higher up in the Santa Rosa Mountains. During winter, they are occasionally seen in the piñon-juniper habitat.

FIGURE 73 Mockingbirds are most common around towns and cities. A few pairs nest in the Colorado Desert's sandy washes.

Mockingbirds are well-known for singing on moonlit nights and for mimicking the songs of other birds, dogs, or even sounds from inanimate objects. Their vocal repertoires are a seemingly endless medley of whistled, squeaky, and harsh notes, each one usually repeated three times.

California Thrasher. *Toxostoma redivivum*
90 grams

Synonyms: Pasadena Thrasher; Southern California Thrasher; Californian Thrasher.

Range: Northern California south to northern Baja California.

Deep Canyon: Resident. Nests in the lower plateau, piñon-juniper, and chaparral habitats.

Thrashers of the genus *Toxostoma* have evolved for a life on the ground. Running is their primary mode of travel, and their wings have become reduced, although they are far from flightless. They forage on the ground, using their curved bills to dig in and whisk through leaf litter and humus for insects. Their musculature has adapted for digging, and their bill curvature is related to the degree of their digging habit (see Engles 1940).

In almost any month of the year, but especially after winter rains, the California Thrasher's song rings out from Deep Canyon's chaparral-covered hillsides. The loud, clear song is made up of separate phrases, some sweet and musical, others harsh. It is frequently given from an exposed perch and tends to be repeated several times. Like many members of the family Mimidae, California Thrashers mimic the songs of other birds, but much less frequently than do Mockingbirds.

California Thrashers are shy and are often seen running between bushes, tails tilted up, as they retreat from intruders. They respond quite well to "squeaking," however, and can usually be coaxed out into the open.

The breeding season is prolonged, beginning in mid-December in parts of the range and lasting until fall (Sargent 1940). At Deep Canyon, most nesting attempts seem to occur in spring. The nest is a cup of coarse twigs lined with rootlets, plant fibers, and grasses. The usual clutch is three, and both parents tend the young (Harrison 1978).

Le Conte's Thrasher. *Toxostoma lecontei*
60–70 grams

Synonyms: Le Conte Thrasher; Gila LeConte Thrasher.

Range: Semiarid and desert regions from central California east to central Arizona and south to northwestern Mexico.

Deep Canyon: Very rare resident of the valley floor and lower alluvial plain.

Le Conte's Thrashers occur in sandy washes and open, flat desert regions in which the soil is fine alluvium or sand. They prefer creosote bush flats that contain cholla for nest sites. They forage for insects in the scanty leaf litter beneath creosote bushes, using their curved bills to whisk and dig in the sand.

Le Conte's Thrashers are secretive. If approached, they flee on foot or in low flight, keeping the thickest cover between themselves and the observer. They usually are seen skulking behind bushes or running in the distance with their black tails cocked up.

Crissal Thrasher. *Toxostoma dorsale*
55–70 grams

Synonym: Arizona Crissal Thrasher.

Range: Southwestern United States south to northeastern Baja California and northwestern and south-central Mexico.

Deep Canyon: Rare summer visitor on the alluvial plain.

Deep Canyon lies at the northwestern edge of this species' range, and the local population is sparse at best. Farther south in the Coachella Valley, as at Indio, Thermal, and Mecca, Crissal Thrashers inhabit dense mesquite thickets (Grinnell and Miller 1944). They have not been found in similar situations on the Deep Canyon Transect.

Where they occur, Crissal Thrashers are usually year-round residents. At Deep Canyon, however, they were found only during the summer and, even then, only in some years. In 1979, at least four thrashers spent the summer in the lower sections of Deep Canyon and Rubble Canyon Washes. A pair that nested in Rubble Canyon Wash had eggs in mid-March and successfully raised two young (see colorplate). By July, all the thrashers had left the area. The following year, none could be found in the same location.

Crissal Thrashers are masters at remaining hidden from view. They forage in and travel through runways beneath dense vegetation. The easiest way to locate them is by their call, a loud *toit-toit* given at dawn and dusk.

THRUSHES—TURDIDAE

Thrushes are a diverse, nearly world-wide family. Over 300 species have been described, and the type genus, *Turdus* (which includes our American Robin), is one of the largest among birds. Many thrushes are superb singers, the European Blackbird (*Turdus merula*) and Nightingale (*Luscinia megarhynchos*) being examples. Most thrushes are terrestrial or partly terrestrial. Many, like the American Robin, forage by hopping across open meadows. The young of most species have spotted breasts.

American Robin. *Turdus migratorius*
80 grams

Synonyms: Robin; Western Robin; *Merula migratorius.*

Range: From treeline in North America south to southern Mexico.

Deep Canyon: Summer visitor in the coniferous forest, and widely scattered winter visitor.

The American Robin requires breeding areas of open ground or turf that provide moist, soft soil. These conditions provide an abundant supply of earthworms or insects that may be caught by shallow probing and mud for the nest cup. The meadows and moist streamsides of Deep Canyon's coniferous forest meet these requisites, and a few pairs of robins nest there.

During winter, robins appear on lawns and in the more luxuriant regions of the valley floor. They come from breeding grounds to the north in the Rocky Mountain and inner coastal districts or from the nearby Great Basin and California mountains (Miller and Stebbins 1964). The number of winter visitors fluctuates widely from year to year.

Western Bluebird. *Sialia mexicana*
27 grams

Synonym: Western Mexican Bluebird.

Range: From southern British Columbia and central Montana south through the mountains of south-central Mexico.

Deep Canyon: Common summer visitor in the coniferous forest. In winter, common in the piñon-juniper woodland, but sparse in the lowlands.

Bluebirds avoid the dry, desert slopes of the Deep Canyon Transect, preferring, instead, regions with denser vegetation. They are most abundant in the piñon-juniper and coniferous forest habitats. A seasonally shifting pattern of habitat preference occurs, with the piñon-juniper being occupied in winter and the coniferous forest in summer (table 42).

The Santa Rosa Mountains' jeffrey pine forest is ideal bluebird breeding habitat. Its open, well-spaced, broken timber provides both nest sites and an abundance of exposed lookout posts at low to middle heights. Not surprisingly, the Western Bluebird is the second most abundant summer visitor in the forest, with a mean density of 28.3 birds/40 ha (see table 38). Bluebirds arrive in the forest in mid-April and depart by early November. They nest in old woodpecker holes, associating and living amicably with swallows, chickadees, and nuthatches in the same large jeffrey pines.

Bluebirds are sit-and-wait foragers. They search open patches of grass, meadow, or even rocky ground from a low perch, flying out to snap

TABLE 42 Seasonal Variation in Western Bluebird Density at Deep Canyon

Habitat	Birds/40 ha			
	Winter	Spring	Summer	Fall
Valley floor	2.2	0	0	0
Piñon-juniper	49.5	10.0	0	8.1
Chaparral	0	1.1	1.6	7.3
Coniferous forest	0	30.7	26.0	25.1

up insects from the ground or grass tops. Occasionally they search larger areas by hovering and take some insects by flycatching.

In winter, bluebirds feed heavily on mistletoe berries. Grinnell and Miller (1944) thought that the presence of this plant might govern the bird's local occurrence. At Deep Canyon, however, other factors are clearly involved, as the abundant mistletoe berries of the dry desert washes rarely attract bluebirds to that habitat.

Townsend's Solitaire. *Myadestes townsendi*
32 grams

Synonyms: Townsend Solitaire; *Myadestes townsendii.*

Range: From Alaska east to southern Mackenzie and thence south to the mountains of southern California and Durango, Mexico.

Deep Canyon: Sparse summer visitor in the coniferous forest, winter visitor in the lowlands.

Audubon named this bird in honor of John Kirk Townsend (1809–1851), one of the notable Philadelphia ornithologists of the first half of the nineteenth century. Townsend (at age twenty-five) discovered the bird in the Pacific Northwest while on a collecting trip with Nuttall. Townsend also discovered and named many western warblers, including Townsend's Warbler, during his brief career.

A Townsend's Solitaire nest, found by Barbara Carlson on an earthen bank near Garnet Queen Creek, is the only definite evidence that this species breeds on the Deep Canyon Transect. They are evidently very rare in the coniferous forest and were encountered on only a few occasions. A period of winter residence from early September to mid-April is to be expected (Miller and Stebbins 1964). Townsend's Solitaires are fond of juniper berries during winter. Thus, many more winter sightings in the piñon-juniper habitat are likely.

OLD WORLD WARBLERS—SYLVIIDAE

The Sylviidae, one of the largest passerine families, contains nearly 400 species in sixty genera. Sylviids are virtually restricted to the Old World (half nest in Africa), with only five species breeding in North America. Most are small, active, insectivorous birds. They are more drably colored than North American wood warblers and are distinguished from them by having ten primary feathers instead of nine.

Mead (1978) describes an important interaction between humans and sylviids: "Bearing in mind their small size, it is astonishing that members of the Sylviidae should feature as human food, but migrants are still caught in some Mediterranean countries for consumption. Cyprus is a particularly bad example and, since there are far flung expatriate Cypriot communities in many countries, there is an international trade in 'pickled

PLATE 22
Gray Flycatcher
Empidonax wrightii

PLATE 23 Screech Owl *Otus asio*

PLATE 24 Swainson's Thrush *Catharus ustulata*

PLATE 25
Western Bluebird
Sialia mexicana

PLATE 26 Townsend's Warbler *Dendroica townsendi*

PLATE 27
Mountain Chickadee
Parus gambeli

PLATE 28 Western Tanager *Piranga ludoviciana*

birds' taken on migration with lime-sticks—these are even imported into Britain and can be bought in London."

Blue-gray Gnatcatcher. *Polioptila caerulea*
5.5 grams

Synonyms: Plumbeous Gnatcatcher; Western Gnatcatcher; Western Blue-gray Gnatcatcher; *Polioptila plumbea.*

Range: Breeds from northern California eastward to the southern Great Lakes region and New Jersey south to the Gulf Coast and the Bahama Islands, and through Mexico to Guatemala. Winters from the southern United States southward.

Deep Canyon: Resident in chaparral and piñon-juniper habitats, winter visitor at lower elevations.

Gnatcatchers were so named for their habit of capturing small insects on the wing. More commonly, they search for insects on the leaves and twigs of trees and shrubs. Occasionally, they pick off insects while hovering like a kinglet. They constantly twitch their tails from side to side while foraging.

This species is easily confused with the Black-tailed Gnatcatcher, especially from August through February when both wear similar winter plumages. During the breeding season, the males can be distinguished as follows: the Blue-gray Gnatcatcher has the black on the head limited to a narrow crescent that extends across the forehead and above the eye (top of the head is *bluish gray*); the Black-tailed Gnatcatcher has a full, black cap and white eye-ring. Female Black-tailed Gnatcatchers (and winter males) have grayish heads. Seen from behind, the *spread* tail shows much more white in the Blue-gray Gnatcatcher in all seasons. White on the Black-tailed Gnatcatcher's tail is limited to the outer margin of the two outer feathers. When *not* fully spread, the Black-tailed Gnatcatcher's tail shows considerable white. The surest way to distinguish the two species is by their calls. The black-tail "sings" a series of harsh, scolding, wrenlike notes, *chee-chee-chee*, quite unlike the plaintive, nasal *peee* of the blue-gray.

Blue-gray Gnatcatchers are most common at Deep Canyon in winter, when migrants from the north arrive on the Colorado Desert. They co-occur with Black-tailed Gnatcatchers on the alluvial plain and valley floor habitats at that season. During summer, they are quite rare, with only a few pairs nesting in the piñion-juniper and chaparral habitats.

Black-tailed Gnatcatcher. *Polioptila melanura*
4.5–6.2 grams

Synonyms: California Black-tailed Gnatcatcher; *Polioptila californica*; Plumbeous Gnatcatcher; *Polioptila plumbea.*

Range: From southern California deserts southeastward to the lower Rio Grande Valley in Texas, thence south to Baja California and central Mexico.

Deep Canyon: Permanent resident.

The Black-tailed Gnatcatcher, a true desert bird, lives year-round in some of the hottest and driest regions of North America. It prefers desert washes grown to cat's claw, palo verde, and smoke tree but can also be found on creosote bush flats. Its preference for deserts is remarkable because its small body size—only slightly larger than a Costa's Hummingbird—necessarily means a high rate of water loss (see Bartholomew and Cade 1963). Rather than being able to drink nectar like the hummingbird, the gnatcatcher must get by on the water contained in its insect prey (insects are about two-thirds water by mass). Unless gnatcatchers are capable of unusual adjustments and are tolerant of high temperatures, their survival during severe heat waves must be tenuous.

Gnatcatchers take a wide variety of insect prey and, considering their size, are capable of predatory acts rivaling those of eagles. I once saw a Black-tailed Gnatcatcher capture and consume a caterpillar twice as thick as its bill and as long as its body. It seized the "beast" with its bill, carried it to a nearby bush, and killed the writhing "monster" by repeatedly thrashing it against a branch. After working over the caterpillar from end-to-end, softening it with numerous bites, the gnatcatcher swallowed it whole and immediately resumed foraging. From my vantage point 3 m away, I wondered if many falconers would be as captivated by this sight as I was. Surely a Golden Eagle killing a snake is merely the same drama played on a larger stage.

Black-tailed Gnatcatchers apparently mate for life, as pairs remain together throughout winter. The breeding season begins in mid-March with construction of the nest, a deep, compact cup of plant fibers and grasses bound together with spider webs (see colorplate). At Deep Canyon, nests are often placed 1 to 2 m above ground in a smoke tree or palo verde. The usual clutch of four eggs is incubated by both parents and hatches in about two weeks. The young fledge nine to fifteen days after hatching but continue to be fed for about three weeks (Harrison 1978).

Gnatcatchers are double-brooded, and sometimes the second clutch is laid in the old nest. This second clutch is usually laid in late May when the weather is hot. Because nests are often located in exposed situations, heat stress is a potential problem. Gnatcatchers are much too small to remain on a nest fully exposed to the midday sun, as Mourning Doves often do, and must abandon the eggs. What prevents the eggs from over-heating while the parent is off the nest? The answer may be the nest's unique structure. Not only is the nest cup deep but it narrows toward the top. Thus, even at midday, the eggs should be well shaded.

Ruby-crowned Kinglet. *Regulus calendula*
6.8 grams

Synonyms: Ashy Kinglet; Western Ruby-crowned Kinglet.

Range: From northwestern Alaska, northwestern Mackenzie, and southeastern Canada south to Guatemala and the Gulf Coast of the United States.

Deep Canyon: Widespread winter visitor.

Kinglets are small, active, unsuspicious birds. They pay little attention to nearby human observers and continuously flick their wings as they flit through the trees in search of insects.

Ruby-crowned Kinglets (fig. 74) begin arriving at Deep Canyon in late September, after the heat of summer has passed. From then until mid-April, they can be found in the valley floor, desert washes, piñon-juniper, and chaparral habitats, foraging alone or occasionally in the company of other small birds. In spring, they depart for their breeding grounds in coniferous forests. Breeding populations occur in the San Bernardino and San Jacinto Mountains. It seems likely that a few pairs nest in the Santa Rosa Mountains, but as yet, direct evidence is lacking.

FIGURE 74 Ruby-crowned Kinglets are common winter visitors at Deep Canyon.

WAGTAILS AND PIPITS—MOTACILLIDAE

A misapprehension, dating back to the fourteenth century, exists concerning the Latin name of this family (see Gruson 1972). Motacillidae does not, as a great many ornithologists believe, mean wagtail but rather "little mover." Since many motacillids habitually pump ("wag") the tail up and down, this error is likely to persist for another 600 years.

The only representative of this family recorded from Deep Canyon is the Water Pipit. From October through April, a few pipits can be seen on the valley floor and alluvial plain, bobbing and bouncing their way over the desert, foraging for insects.

WAXWINGS—BOMBYCILLIDAE

The sleek, silky-plumaged waxwings are attractive birds, dressed in somber grays and browns, with a yellow tail band. A prominent crest accents their heads, and a red drop of "wax," which tips the secondary (innerwing) flight feathers, gives the family its vernacular name. The voice of waxwings is a high, thin lisp or *zeee*, sometimes slightly trilled.

Waxwings occur in flocks, except during the breeding season, and frequent wooded habitats and surburban areas. During winter, they are conspicuous at berry sources such as ornamental shrubs (especially pyracantha), mistletoe, and toyon (*Heteromeles*). They are often found in the company of other berry-eating species such as American Robins. Waxwings are noted for their irruptions and irregular winter occurrences: they may be abundant in some years, while distinctly uncommon in others. Winter invasions in southerly regions are believed to coincide with low fruit yields and high bird numbers in the north.

To date, only the Cedar Waxwing has been seen at Deep Canyon. A few were found in spring in the piñon-juniper habitat, on the alluvial plain, and around urban gardens. Thus, they are mainly migrants or late winter visitors on the transect. They should occur throughout the winter in some years, however. Cedar Waxwings breed far to the north of Deep Canyon, from southeastern Alaska south to extreme northwestern California.

SILKY FLYCATCHERS—PTILOGONATIDAE

This small group (four species) of tropical American fruit- and insect-eaters inhabits dry, brushy country from the southwestern United States south to central Panama. Like their close relatives, the waxwings, they have soft plumage and prominent crests.

Phainopepla. *Phainopepla nitens*
22–28 grams

Synonym: Northern Phainopepla.

Range: From central California southeastward to western Texas, south through Baja California and the Mexican Plateau to Puebla and Veracruz.

Deep Canyon: Present in three roles: common winter visitor, summer visitor, and migrant.

Phainopeplas are the northernmost representative of the Ptilogonatidae. Their habit of sallying forth from high perches in buoyantly languid flight makes them one of the most conspicuous of the Colorado Desert's birds. Their generic name is derived from the Greek words for "shining robe" (from *phaeinos*, "shining," and *peplos*, "robe"), an allusion to the silky, shining plumage of the male. Males are glossy black with a long, pointed crest and conspicuous white "windows" in the wing (seen in flight). Females resemble males but are a duller, dark gray in color.

In North America, Phainopeplas winter on the Sonoran Desert and disperse to regions east, west, and north in spring. The first males arrive at Deep Canyon in October and soon establish territories in the desert washes or around the mesquite dunes of the valley floor, where they are closely associated with desert mistletoe clumps. Throughout their principal period of residence (October through May), they are most abundant on the alluvial plain and in the piñon-juniper habitat (fig. 75).

Mistletoe berries (fig. 76) are the adult Phainopepla's major food on the Colorado Desert, and a mutualistic relationship has evolved between berry and bird (Walsberg 1977). The desert mistletoe, a common parasite

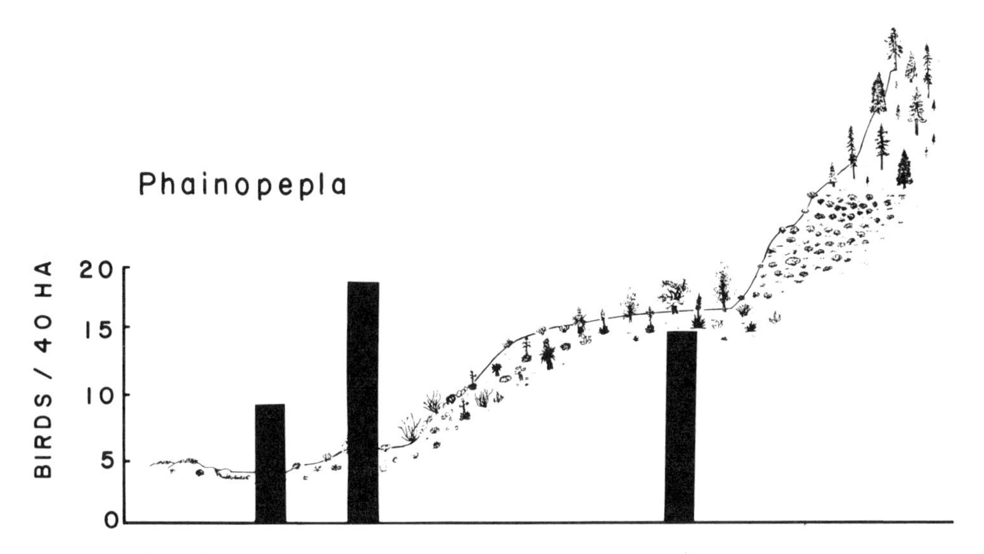

FIGURE 75 Mean winter density of Phainopeplas at Deep Canyon. See fig. 38 for explanation of symbols.

FIGURE 76 Desert mistletoe berries, abundant from November through March, are eaten by many desert birds.

of several arborescent legumes, depends on Phainopeplas, Mockingbirds, and a few other species to disperse its seeds. Seeds passed in the feces accumulate on plant branches and, after germinating and penetrating the bark, establish new mistletoe clumps. In a sense, Phainopeplas "plant" a crop for future harvest.

Desert mistletoe at Deep Canyon produces berries mainly between November and April. Berry abundance peaks in December or January and declines steadily thereafter (see Walsberg 1977). But flying insects—the Phainopepla's other major food item—do not peak until summer (see fig. 22). Berries constitute the bulk of the adult's diet, whereas nestlings are fed both insects and berries. In some years, sufficient fruit and insects to permit breeding exist for only a six- to eight-week period. Phainopeplas require five weeks from clutch completion to fledging, and their social system exhibits several adaptations to a compressed and semipredictable breeding season (see Walsberg 1977). These include nest building by the unmated males and early initiation of courtship (in some areas as early as January).

The Phainopepla's nest, a shallow cup saddled on a horizontal limb (fig. 77), is typically placed 2 to 3 m above ground in mesquite or palo verde trees. On the Colorado Desert, the usual clutch is two. Although typically both parents incubate the eggs and tend the young, I found

FIGURE 77 Phainopepla nest in a palo verde tree.

several nests tended exclusively by males. Whether their mates had been killed by predators or merely had abandoned the area is unknown.

Most Phainopeplas leave Deep Canyon in April and May. Their destination is uncertain, but Walsberg (1977) suggests it is the coastal woodlands of the Pacific slope (fig. 78), where they remate and breed again in June and July. Before leaving Deep Canyon, loose flocks of ten or more adults and juveniles gather at large mistletoe clumps. Agonistic behavior, conspicuous during the winter, abates at this time. This may be a prelude to the change in social behavior that accompanies the move to the coastal woodlands; for, in contrast to their behavior on the desert, Phainopeplas in the woodlands engage in social feeding, nest in colonies, do not establish large feeding territories, and seldom engage in display flights (see Walsberg 1977). Such a dramatic change in the social system within a single season is remarkable among birds. How such a system could have evolved is difficult to imagine, and additional studies on this fascinating bird are warranted.

In May, while most Phainopeplas are leaving Deep Canyon, a few individuals, presumably migrants from farther south, arrive in the chaparral and piñon-juniper habitats, where they breed in June and July. On 28 June 1979, I found a Phainopepla nest in a redshank on the chaparral slope near the spot where Deep Canyon Creek crosses the jeep road that

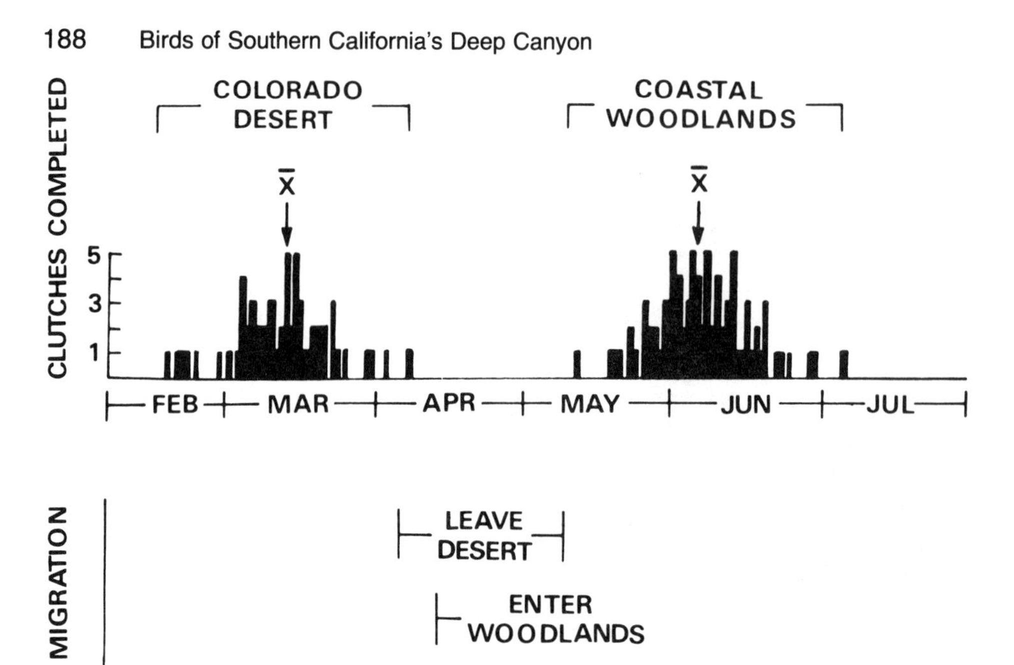

FIGURE 78 Timing of egg laying and migration in the Phainopepla's dual breeding ranges. Nesting occurs first on the Colorado Desert, then in California's coastal chaparral (figure from Walsberg 1977).

ascends Santa Rosa Mountains' north slope. The nest contained a five- to seven-day-old chick, and both parents were nearby.

During the time that most Phainopeplas are on the Colorado Desert (October to May), temperatures are mild. This, and the high water content of their diet, frees Phainopeplas from the water-balance problems that summer residents face. Water turnover of Phainopeplas in late April and early June averages nearly 95 percent of body mass per day (Weathers and Nagy 1980). This high rate suggests that Phainopeplas might have difficulty maintaining water balance at high temperatures and may explain why Phainopeplas breeding in Deep Canyon's chaparral in summer were stationed near water.

The most frequently heard sounds of the Phainopepla are a single, soft, low *quirt* and a harsher *churr*. Because their song is variously described as weak, formless, and disconnected (Peterson 1961, Hoffmann 1955), I was surprised to find that Phainopeplas are excellent vocal mimics. Furthermore, they seem to use mimicry to distract attacking predators. Several North American passerines mimic the vocalizations of other species, the Mockingbird being a classic example. Most reports of passerine mimicry, however, concern species that occur outside of this country (Borror 1977). To my knowledge, no reports of mimicry by the Phainopepla exist.

During April 1979, my wife and I observed mimicry in seven Phainopeplas (three males, three females, and one juvenile of unknown sex). All produced soft, muted calls in response to handling during removal from mist nets. Although we were not equipped to make sound recordings of the calls, the imitations were excellent and the calls mimicked were so distinctive that identification was easy. Syllabic representations of imitated calls are as follows: "*keeer-r-r*"—Red-tailed Hawk; "*yuk-kwair'go-o*" and "*kway-er*"—Gambel's Quail; alarm-scold "*see-lip*" of the Verdin; and "*churr-churr-churr-churr*"—Cactus Wren. All of the species mimicked are common in the study area.

Phainopeplas seldom mimic in the wild. But, on 25 April, my wife heard a female Phainopepla give the Red-tailed Hawk's "*keeer-r-r*" cry, just as it was attacked by a Loggerhead Shrike. This suggests that the call might function as a distraction. In this case, however, the undistracted shrike killed the Phainopepla. When we released a trained shrike in a room containing caged Phainopeplas, they immediately responded by giving the Red-tailed Hawk's call. Colorado Desert shrikes prey on both adult and nestling Phainopeplas. Because the prey of Red-tailed Hawks includes birds the size of shrikes (Bent 1937), mimicry of the "*keeer-r-r*" call might distract attacking shrikes and thus have survival value.

SHRIKES—LANIIDAE

Shrikes are primarily an Old World group of seventy-four species of carnivorous songbirds. They are most numerous in tropical Africa, with only two species reaching North America—the Northern Shrike (*Lanius excubitor*) and the Loggerhead Shrike. All are bold and aggressive predators of small birds, mammals, reptiles, and insects. They kill with a swift, firm bite.

Loggerhead Shrike. *Lanius ludovicianus*
45–55 grams

Synonyms: California Shrike; California Loggerhead Shrike.

Range: From southern Canada south through Mexico. Winters southward from 45° north latitude.

Deep Canyon: Sparse but widespread resident.

Loggerhead Shrikes require open terrain with well-spaced lookout posts from which to scan the ground and air for prey. Consequently, shrikes are most abundant at Deep Canyon's lower elevations. They reach their greatest density on the open, rocky slopes habitat (fig. 79), where they are most often seen perched atop ocotillo and palo verde trees.

Shrikes are known as "butcher birds" for their habit of impaling their prey on thorns. They do this mainly with larger animals, such as rodents and birds, because their feet are too weak to hold the prey while it is torn apart. Small prey are held in one foot and torn apart or swallowed whole. Shrikes are not, as is sometimes claimed, wanton killers, since they usually return later to consume the impaled remains. They have favorite impal-

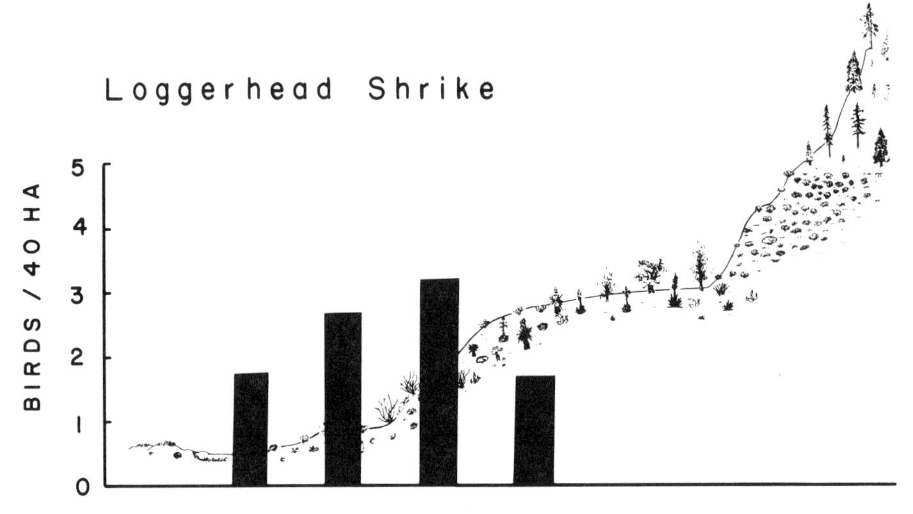

FIGURE 79 Mean annual density of Loggerhead Shrikes at Deep Canyon. See fig. 38 for explanation of symbols.

ing stations to which they consistently return. At Deep Canyon, impaling stations are often located within a meter of the ground in dead creosote bushes. During the breeding season, males frequently impale prey near the nest tree for the female's use.

Shrikes are solitary, except during the breeding season, and maintain territories throughout the year. On the Mojave Desert, shrike territories are on the order of 14 to 18 ha (Miller 1931). A 350-ha section of Deep Canyon's alluvial plain (located below the Boyd Research Center and east of Deep Canyon Wash) supported 9 shrike pairs, or 1 shrike per 20 ha, close to Miller's value.

The following description of the Loggerhead Shrike's annual cycle is based on my own observations and those of Miller (1931). In early July, at the close of the breeding season, the family groups begin to disintegrate. The young still follow the adults and beg but are no longer fed. They steal food from their parents whenever possible, and gradually the adults become intolerant of the young and drive them off. By August, the family groups dissolve entirely, including the adult pair, and each individual seeks an exclusive feeding territory. Frequently, the male and female occupy adjacent territories, but each hunts independently of the other and maintains a separate feeding area.

Throughout fall, when shrikes are solitary, intruders are attacked, chased, and scolded with a loud *bzeek* call that is sounded several times. This call and a series of rapid staccato notes indicate extreme agitation. Shrikes are more vocal during fall and winter than at other times of year and "sing" daily at first light. The dawn song is a rhythmic repetition of several different notes, each one repeated about once a second for a minute or more before another note is sung. Commonly heard notes include *tweedle-dee, chee-chunck, aaayh,* and several screeches and clear notes that are difficult to represent phonetically. Although Loggerhead Shrikes are sometimes claimed to mimic the calls of other species, in my experience (and that of Miller) they do not.

I have not observed courtship in shrikes, but according to Miller it "chiefly consists of certain notes of excitement, characteristic nuptial flights, and sexual posturing." At Deep Canyon, pairing occurs as early as November or as late as February, depending on the pair. During the early stages of the breeding cycle, females beg from and are fed by their mates, but both members of the pair spend considerable time foraging alone.

The nest, built by the female alone, is a loose, bulky bowl of twigs and sticks lined with a thick, felted cup. Nest construction at Deep Canyon occurs between late February and early April, with most nests placed about 1.5 to 2 m above ground in palo verde trees or in large clumps of mistletoe. Frequently, the same nest tree is used in successive seasons, and occasionally a new nest is built atop an old one.

The usual clutch of four to six eggs is incubated by the female alone. Incubation starts with the next-to-the-last egg. Thus, one chick is always smaller than the others. During incubation, the male brings food to the female or sometimes hangs it in a nearby tree. Females may leave the nest to forage on their own, but in the early stages of incubation they remain on the nest 80 percent of the day.

The eggs hatch in sixteen days. During the early nestling period, the female broods the young almost continuously and is fed on the nest by her mate. As the young grow, the male continues to bring food to the nest. Usually he gives it to the female, who in turn feeds the young. The male will feed the young during the female's absence, however. After the young are about one week old, the female joins the male in foraging for prey. As the young approach fledging age, they perch on the nest's edge or in the nearby branches (fig. 80) and begin to stretch and flap their wings. Between age 21 to 25 days, the young are usually in the nest tree but not in the nest. After the 25th day, the young follow their parents about, begging for food with quivering wings and an incessant whine. By about the 36th to 40th day after hatching, the young are feeding on their own, and the adults may begin a second brood.

Loggerhead Shrikes show wide latitude in the type of prey sought. They eat virtually every kind of insect, including bees and obnoxious-smelling beetles, and any birds and rodents that they can kill. The majority of birds taken are small, but I have often seen shrikes pursue Mourning Doves (once even a Gambel's Quail), although they never actually killed one. Northern Shrikes are able to kill large birds—Mourning Doves, Hairy Woodpeckers, and American Robins (see Miller 1931, p. 198)—and I believe Loggerhead Shrikes occasionally succeed in killing Mourning Doves, a bird more than twice their size.

Shrikes resemble accipiter hawks in having long, broad tails and short, rounded wings. Like accipiters, they are adept at pursuing prey by flying through brush or by running along the ground. Shrikes are amaz-

FIGURE 80 A few days before fledging, young Loggerhead Shrikes leave the nest to roost in the nest tree.

ingly agile and quick. One I saw pursuing a fledgling House Finch reminded me of a flycatcher chasing a moth: wheeling, turning, and diving after its quarry with snapping bill.

Shrikes prey on nestling birds as well as adults and can have a catastrophic effect on the nesting success of other species (e.g., Reynolds 1979). One Deep Canyon shrike pair, with a brood of four nestlings, was observed foraging in Coyote Wash during the spring of 1979. In an eight-hour period, they took four nestling Phainopeplas, a brooding adult Black-tailed Gnatcatcher, four House Finch nestlings, and a brood of Verdins. Observations of shrike raids on nests were memorable experiences. Noisy flocks of mobbing Phainopeplas, Mockingbirds, Black-tailed Gnatcatchers, and Verdins swarmed nearby but were powerless to stop the shrikes, as screaming chicks were torn from the nest. Subsequent observations at other shrike nests revealed this to be an extreme example, reflecting the high concentration of nests in these shrikes' territory. Other pairs foraged mainly on reptiles and insects, taking fewer nestlings. Nevertheless, the effect of shrikes on bird communities can be striking.

On the Colorado Desert, Loggerhead Shrikes are able to survive without drinking water. They will, however, come to water if it is available, both to drink and to bathe.

STARLINGS—STURNIDAE

This Old World family of over 100 species ranges throughout Europe, Asia, Malaya, and Australia. Many species are black or predominantly black in color, often with considerable iridescence. Several, like the familiar starlings and mynahs, are excellent mimics of other birds' songs. Most nest in holes in trees, and some (e.g., Starlings) have secondarily adapted to nesting in holes in rocks, buildings, and nest boxes.

Starling. *Sturnus vulgaris*
75 grams

Synonym: European Starling.

Range: Introduced in North America; widespread throughout much of Canada and the United States.

Deep Canyon: Permanent resident near towns.

Starlings, much maligned by North American bird watchers, are not native to North America. They are extremely aggressive (frequently driving out the colorful native birds by taking over their nest sites), inhabit colonial roosts that are filthy messes, and have multiplied from an original handful of immigrants to the present teeming millions. In all these respects, they are no different from us humans, and they at least have done considerably less damage to the environment.

On the Deep Canyon Transect, Starlings are most commonly seen in towns. They also occur in the valley floor's creosote bush scrub, however, where they are the second most abundant resident (see table 8).

VIREOS—VIREONIDAE

Vireos are an exclusively New World group of thirty-eight small, drably colored insectivores. They are more numerous in North and Central America than in South America. North American vireos defend their breeding territories with loud songs and aggressive behavior. Most are migratory and winter in Mexico and Central America.

Typical vireo nests, pensile cups of plant fibers, are suspended by the rim in a thin twig-fork. Incubation and care of the young is by both sexes. Incubating adults sit very tightly and some can even be touched while on the nest.

Bell's Vireo. *Vireo bellii*
8–13 grams

Synonyms: Least Vireo; Least Bell Vireo; Bell Vireo; *Vireo belli.*

Range: Breeds from the Central Valley of California east to northeastern Illinois, thence south to southern Texas and northwestern Louisiana. Winters from northeastern Mexico south to northern Nicaragua.

Deep Canyon: Status uncertain; extremely rare.

In 1908, this species was abundant along streams in the lower canyons of the transect. Grinnell and Swarth (1913) described it as "particularly numerous" in lower Palm Canyon during the middle of June. Since then, it has declined markedly and was placed on the *American Birds* "Blue List" of threatened birds in 1973 (Wilbur 1979). During the present study, Bell's Vireos were found only during April (and thus were probably on migration). I could find none breeding in the canyons. Grinnell and Miller (1944) commented that "in the past fifteen years a noticeable decline in numbers [of Bell's Vireos] has occurred in parts of southern California . . . apparently coincident with [an] increase of cowbirds which heavily parasitize this vireo." Male Bell's Vireos habitually sing on or near the nest, a behavior that cowbirds doubtlessly find helpful. Gray Vireos, in contrast, do not sing on the nest, and their numbers have changed little since 1908.

Gray Vireo. *Vireo vicinior*
12–13 grams

Synonym: California Gray Vireo.

Range: Breeds from southern California, Nevada, Utah, central Arizona, and western Oklahoma south to northwestern Baja California, southern Arizona, and western Texas. Winters chiefly in southern Baja California and Sonora (fig. 81).

Deep Canyon: Summer resident in the chaparral, and migrant.

The Gray Vireo is more tolerant of heat and aridity than other vireos. It actually prefers the hot, dry, chaparral and piñon-juniper woodlands of the southwest. It lives on a diet of insects gleaned from shrubs and never needs to drink water. At Deep Canyon, this vireo is closely associated with chaparral dominated by *Adenostoma.* Grinnell and Swarth (1913) noted several individuals "on a brushy ridge near Garnet Queen Mine" and estimated their density in chaparral to the west of Deep Canyon as 2.5 birds/40 ha. This is close to their present density (1.6/40 ha; see table 35). In the Mojave Desert, Gray Vireos nest in piñon-juniper (Miller and Stebbins 1964), and a few may nest in this habitat at Deep Canyon. But, as yet, no nests have been found there.

Solitary Vireo. *Vireo solitarius*
14–16 grams

Synonyms: Cassin Vireo; Cassin Solitary Vireo; Blue-headed Vireo; *Lanivireo solitarius.*

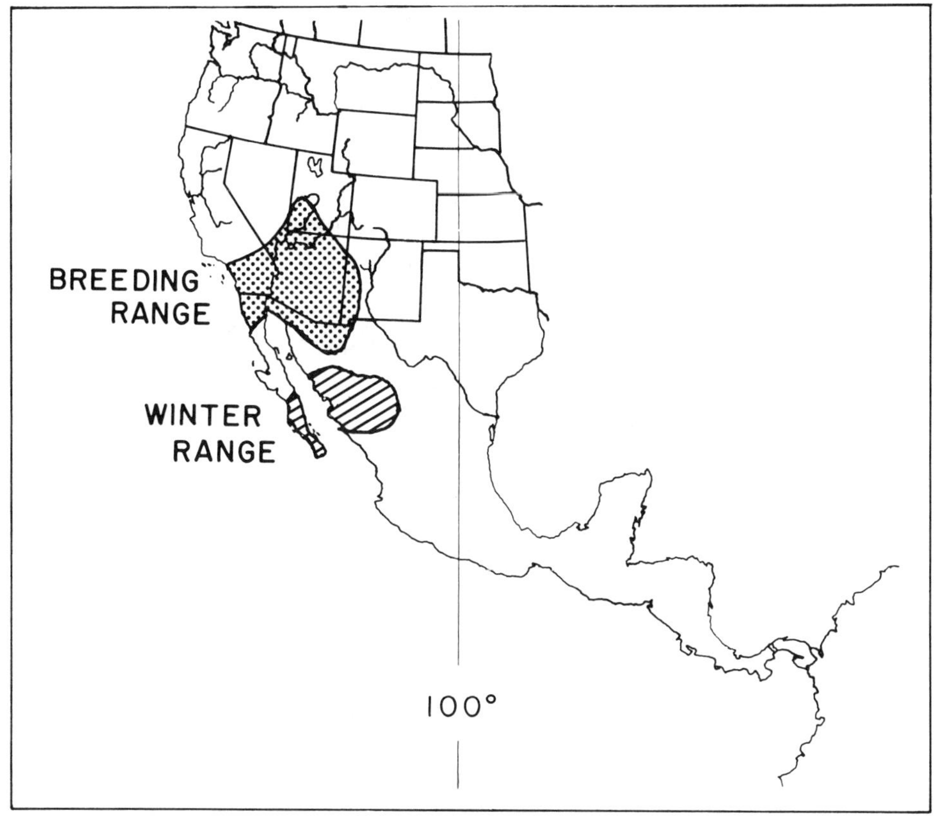

FIGURE 81 Map showing principal breeding and wintering ranges of the Gray Vireo.

Range: Breeds from southern Canada south to El Salvador. Winters from Arizona and South Carolina south through western Mexico to northern Nicaragua and Cuba.

Deep Canyon: Summer visitor and sparse spring migrant.

A few migrating Solitary Vireos pass through Deep Canyon's alluvial plain during April and May. Others establish summer residency and breed along Garnet Queen Creek, where the oak and coniferous forest offers the open branch-work favored by this species. The number of Solitary Vireos at Garnet Queen Creek apparently has declined since 1908, when Grinnell and Swarth (1913) encountered "many" and collected five.

Warbling Vireo. *Vireo gilvus*
10–12 grams

Synonyms: Western Warbling Vireo; *Vireosylva gilva.*

Range: Breeds throughout North America in general, from Great Slave Lake to northern Mexico. Winters from Sonora, Mexico, south to Guatemala and El Salvador.

Deep Canyon: Widespread but sparse migrant; summer visitor in coniferous forest.

Warbling Vireos move through Deep Canyon on spring migration in April and May. Although this species is the most frequently seen vireo, its numbers are still quite modest. During spring censuses of the desert wash habitat, Warbling Vireos were seen only 8 percent of the time (see table 15). This species has been seen in all of Deep Canyon's habitats, from the coniferous forest to the valley floor.

WOOD WARBLERS—PARULIDAE

Wood Warblers comprise a large (119 species) and varied family of mainly small, arboreal insectivores. They are confined to the New World from Alaska south to Argentina. The parulids actively forage among foliage, picking insects with their slender, pointed bills. Many species supplement their diet with nectar and fruit. Males are brilliantly colored and typically show some yellow, red, or chestnut.

Of Deep Canyon's eleven species of warblers, all but three are pure migrants, and most are uncommon. The Hermit Warbler serves to illustrate the occurrence pattern of these migrants (fig. 82). It winters in Mexico and Central America as far south as Nicaragua and breeds in the Pacific Northwest and the Sierra Nevada Mountains. It passes through Deep Canyon on spring migration between mid-April and mid-May and makes the fall return around the end of August.

The spring warbler migration occurs when temperatures on the desert are usually moderate (but sometimes hot), foliage fairly luxuriant, and insects abundant. Fall migration, however, occurs when desert temperatures are considerably higher (see fig. 9) and insect numbers low. Warblers respond to this seasonal difference by traveling different routes during spring and fall. During spring, warblers concentrate in the desert washes. In fall, most warblers move south along the mountain tops and avoid the dry, lower slopes (table 43). Nevertheless, some warblers (mainly yellow-rumped and orange-crowned) travel the desert washes during fall, and they must contend with truly severe conditions. These fall transients must obtain sufficient insects—perhaps more for water than for energy—while foraging in unfamiliar vegetation to which they are not specially adapted. Many warblers perish during the fall transit, and immatures are conspicuous among the casualties. Miller and Stebbins (1964) described the consequences of fall migration through the Mojave Desert as follows: "We found dead birds, dried and emaciated; we saw others, scarcely able to fly, so anxious for food, water, and shade that in the heat of the day they came into the shelter of our camps almost under our feet; and we took specimens in such poor condition that they almost certainly would have succumbed before they could travel the remaining 500 miles across the

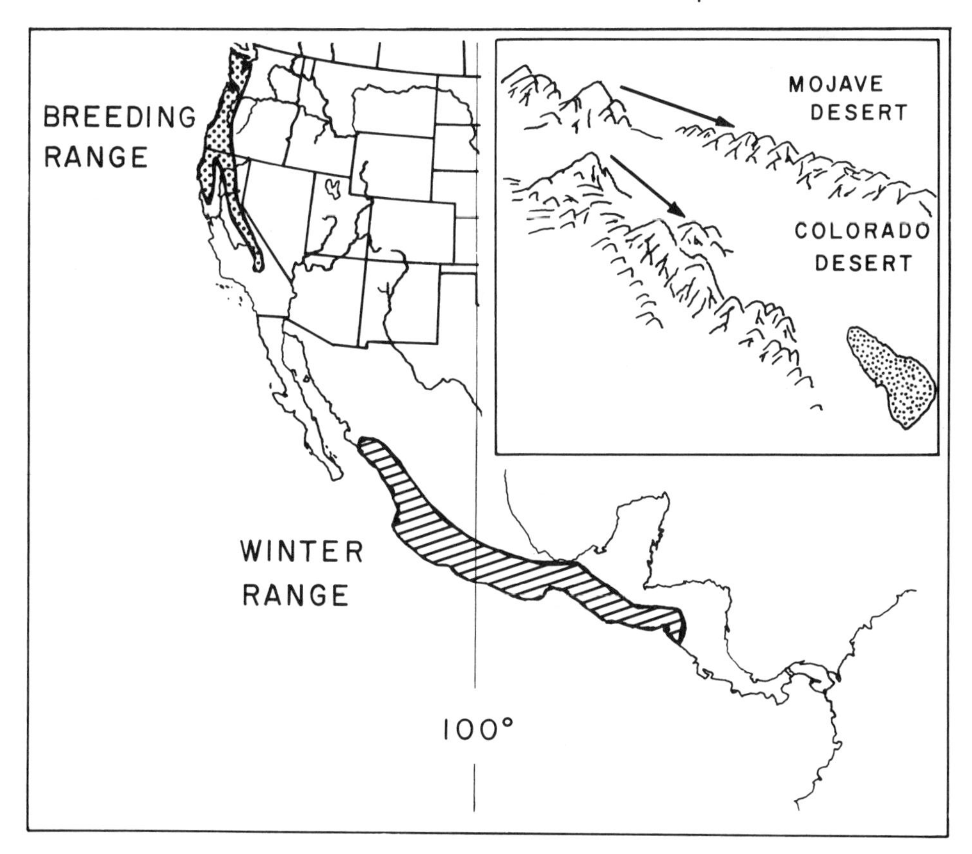

FIGURE 82 Map showing principal breeding and wintering ranges of the Hermit Warbler and some routes (arrows) followed in fall migration. Most warblers seen at Deep Canyon travel south through the mountains in fall, thereby avoiding the hot, dry desert (modified from Miller and Stebbins 1964).

TABLE 43 Seasonal Variation in Warbler Density at Deep Canyon

Habitat	Birds/40 ha	
	Spring	Fall
Valley floor	0	0
Alluvial plain		
Scrubland	0.56	0
Desert wash	18.83	3.19
Rocky slopes	3.25	0
Lower plateau	0.87	0
Piñon-juniper	4.15	0.38
Chaparral	0.53	2.91
Coniferous forest	0.67	9.07

Colorado Desert basin." Thus, warblers of western North America face a problem not encountered by the more numerous warbler species of the eastern United States—migration across deserts. Miller and Stebbins (1964) speculate that this is one reason why there are fewer warbler species in the west.

Warblers usually migrate at night, stopping to rest and feed during the day in vegetation that resembles that of their breeding grounds. Thus, Hermit and Townsend's Warblers, both of which breed in conifers, are seen mostly in Deep Canyon's piñon-juniper and coniferous forest habitats. Wilson's Warblers, however, prefer streamside tangles and forage more often in lower, brushy cover.

Orange-crowned Warbler. *Vermivora celata*
8–9 grams

Synonyms: Lutescent Warbler, Lutescent Orange-crowned Warbler; *Helminthophila celata.*

Range: Breeds from central Alaska southeastward across Canada to northwestern Quebec, thence south to northwestern Baja California and western Texas. Winters from southwestern United States and South Carolina south to Guatemala.

Deep Canyon: Migrant and summer visitor.

This is the most common transient warbler seen in Deep Canyon's desert wash habitat, up to five per hour being seen during the spring migration peak. Extreme spring dates of occurrence are 27 March and 12 May. As a fall migrant, they are much more common in the chaparral than the desert washes.

Orange-crowned Warblers forage individually when on spring migration. They drift through the palo verde trees of the desert washes or through scrubland creosote bushes, eating mostly leaf bugs, caterpillars, aphids, flies, and spiders.

A few pairs nest in Deep Canyon's chaparral near Garnet Queen Creek, where the steep montane slopes and dense cover provide typical breeding habitat. The nest, a bulky cup of coarse grasses, bark strips, and plant down, is typically built on the ground at the base of a bush. Apparently the Orange-crowned Warbler has colonized the Santa Rosa Mountains since 1908, as Grinnell and Swarth (1913) found only one individual during spring and no indication of local nesting.

Yellow-rumped Warbler. *Dendroica coronata*
11–14 grams

Synonyms: Pacific Audubon Warbler; Audubon('s) Warbler; *Dendroica auduboni.*

Range: Breeds from southwestern Canada south through western mountains to the Sierra Madre Occidental in Mexico. Winters from southwestern British Columbia south to Costa Rica.

FIGURE 83 Female Yellow-rumped Warbler in spring plumage.

Deep Canyon: Occurs year-round, but in different roles: winter visitor, summer visitor, and migrant.

The large and hardy Yellow-rumped Warblers winter farther north than other North American wood warblers. Their tolerance of cold winter climates is probably more a function of diet than size. They eat not only insects but also seeds and berries.

The majority of Yellow-rumped Warblers seen at Deep Canyon belong to the Audubon's race (fig. 83). Some arrive on the transect's valley floor habitat in November, spend the winter, and leave in early April. During this period, they can be found (never in great numbers) in the mesquite thickets or foraging among the creosote bushes of the sandy flats. They were detected during 56 percent of the valley floor winter censuses. In contrast, no Yellow-rumped Warblers were detected during the desert wash censuses of December, January, and February, although a few were observed off the census strips. Apparently, this species prefers the warmer valley floor as a wintering spot.

The northward migration begins early, as waves of Yellow-rumped Warblers were observed on the alluvial plain on 11 February 1979—as many as ten individuals in sight at one time. By early May, most of the migrants have passed. From then until their fall return in late September, none are seen at the lower elevations.

A very few pairs breed in the coniferous forest of the Deep Canyon Transect, as they did at the time of Grinnell and Swarth's study in 1908. They arrive there in mid-April and leave by early November. An increase in numbers in the coniferous forest in September corresponds to the period of southern migration.

Black-throated Gray Warbler. *Dendroica nigrescens*
7–9 grams

Synonym: Arizona Black-throated Gray Warbler.

Range: Breeds from British Columbia south to northern Baja California and southern New Mexico. Winters from southern California and Arizona south through southwestern Mexico.

Deep Canyon: Migrant and sparse summer visitor.

In the north, the Black-throated Gray Warbler nests in fir forests. But through most of its breeding range, it prefers warm and moderately dry regions of fairly dense foliage that is often harsh and stiff. On the desert mountains and interior plateaus, the Black-throated Gray is found in golden-cup oaks (especially where intermingled with chaparral), piñon-juniper woodlands, and dry coniferous forests.

A few pairs nest on the Deep Canyon Transect in the dry stands of golden-cup oaks near Garnet Queen Creek. The local population apparently has declined markedly since the early 1900s, as Grinnell and Swarth (1913) described them as "abundant" at Garnet Queen Mine between 25 June to 2 July. Today, they are uncommon and were not detected during several July visits to the area. Furthermore, they were not found in the piñon-juniper habitat during the breeding season.

Black-throated Grays migrate through the Deep Canyon Transect between early April and mid-May. At that time, they occur on the alluvial plain upslope to the coniferous forest. No clear evidence of a fall migration has been seen, but in the Mojave Desert migration occurs from late August through late October (Miller and Stebbins 1964).

WEAVER FINCHES—PLOCEIDAE

The family Ploceidae contains over 300 species of small- to medium-sized, seed-eating passerines. Some colonial species are known for their well-woven nests. Other species nest in natural cavities. Originally confined to the Old World, the family has colonized the world, except for a few oceanic islands. Two introduced species, the House Sparrow and the European Tree Sparrow (*Passer montanus*), are established in North America.

House Sparrow. *Passer domesticus*
26–30 grams

Synonyms: English Sparrow; English House Sparrow.

Range: Introduced in North America; widespread, most common near developed areas.

Deep Canyon: Permanent resident. Closely associated with towns and houses. Seldom wanders away from developed areas.

The following invective was fired by Joe Marshall (Phillips et al. 1964): "This loud and messy bird shares with two other species, *Rattus rattus* and *Homo sapiens*, traits which make the three of them a blight upon the earth: omnivorous food habits, ability to colonize every corner of the world, and an inordinate ability to procreate—in geometric progression." Like the Starling, this species was deliberately introduced into North America by man and has become widespread. Unlike the Starling, it does not compete seriously with native species for nest cavities,preferring instead to nest in and around buildings. At Deep Canyon, House Sparrows are seen most often near towns and seldom venture into undisturbed areas.

Many ecological studies have involved this agricultural pest. Two recent books summarize excellently this body of knowledge (Kendeigh and Pinowski 1973, Pinowski and Kendeigh 1977).

BLACKBIRDS AND ORIOLES—ICTERIDAE

The Icteridae, a diverse Pan-American family noted for its adaptive radiation, includes arboreal, terrestrial, territorial, colonial, and parasitic species. The American tropics is the distribution center of the family's approximately 100 species. Most icterids are sexually dimorphic in color and size, with males being the larger and more brightly colored sex. Black is the predominant family color, but in many species it is relieved by yellow, red, or brown.

Western Meadowlark. *Sturnella neglecta*
Males 95–125 grams; females 75–105 grams

Synonym: *Sturnella magna.*

Range: Southwestern Canada eastward to southern Ontario, thence south to south-central Mexico.

Deep Canyon: Resident; most abundant near golf courses.

Meadowlarks are fairly common residents of the lower elevations of the Deep Canyon Transect. These large, conspicuous birds, bright yellow below with a black "V" on the breast, sit on fence posts and have loud, melodious calls. They forage for insects by walking on the ground and are frequent visitors to lawns and golf courses. A few nests have been found in the dense grass that lines the Ironwood Golf Course's fairways.

Most icterids build hanging, purselike nests knit from grasses or plant strips. The Western Meadowlark's nest, in contrast, is a cup built on the ground in growing grass and concealed by a domed top of overhanging grasses. The typical clutch is five to seven. Incubating females sit very tightly, usually not flushing from the nest until nearly stepped on.

Meadlowlarks seen on the valley floor from late October through February are winter visitors from the northern portions of the species' range. Wintering birds are gregarious and occur in flocks.

FIGURE 84 Male Hooded Oriole at an ocotillo blossom.

Hooded Oriole. *Icterus cucullatus*
Males 23–28 grams; females 21–25 grams

Synonyms: Arizona Hooded Oriole; California Hooded Oriole.

Range: Southern United States from central California to southern Texas, thence south through Mexico to Honduras. Winters from southern California (casually) southward.

Deep Canyon: Migrant and summer visitor.

Hooded Orioles are among the most attractive of Deep Canyon's birds. Adult males (fig. 84) have solid black tails, black wings with white patches, and orange bodies accented by a black bib and back-patch. Females are dull olive-green above and yellowish below. Males attain their full plumage in their second year and, except for partial development of the black bib, resemble females during their first year. Development of the male's showy plumage is not a requisite for breeding, however, as some first-year males nest (Phillips et al. 1964).

Migrating Hooded Orioles begin arriving on Deep Canyon's alluvial plain in late March. If enough rain fell the preceding winter to trigger an ocotillo bloom, the orioles will be seen at the crimson, waxy blossoms inserting their long, thin bills down the tubular flowers to obtain nectar, thereby effecting pollination. For the next month, the number of Hooded

Orioles on the alluvial plain fluctuates from zero to four or five, as migrants arrive and leave. By late April, most migrants have departed, and summer visitors have left the alluvial plain for their breeding stations up the canyon.

Hooded Orioles nest in the fan palms that line Deep Canyon's side canyons, as well as in those planted on golf courses or around homes. The stringy palm fibers provide ideal building material for the oriole's pouch-like nest. The female sews the nest fibers securely through the palm leaves, anchoring the nest against the lashing desert winds. Nests are typically placed 3 to 15 m above ground, frequently within the palm's "skirts." Such sites are well shaded and protected from nonflying predators, but Scrub Jays easily find and raid oriole nests.

Scott's Oriole. *Icterus parisorum*
32–42 grams

Synonym: Scott Oriole.

Range: Breeds from southeastern California and southern Nevada east to western Texas and south to southwestern and central Mexico. Winters mainly from northern Baja California and Sonora southward to Oaxaca (fig. 85).

Deep Canyon: Chiefly a summer visitor, but a few occur in winter.

This bird was named *Icterus scottii* by Lt. Couch for Gen. Winfield Scott (1786–1866), an associate of every American president from Jefferson to Lincoln and a political force in the United States. Later it was found that Bonaparte had previously named it in honor of the Paris brothers of France—owners of an early ornithological collection.

The adult male Scott's Oriole is clear yellow and black without any orange tints. His song, a series of loud, clear, whistled phrases, suggests that of the Western Meadowlark. Scott's Orioles sing frequently during April and May, often from a high perch, and are thus very conspicuous. In June they become silent, abandon their singing posts, and seem much less numerous.

Scott's Orioles begin arriving at Deep Canyon in moderate numbers in early March. Although they are widely distributed throughout the transect at that time, they are most conspicuous in the desert washes and near stands of blooming ocotillo. On 2 April 1979, I counted eight Scott's Orioles during a one-hour walk in Sheep and Coyote Washes. This was an all-time high: usually only one to three orioles are seen during a one-hour walk. Many of the orioles found in the desert washes in March and April are migrants.

The diet of Scott's Orioles consists of insects, fruit, and nectar. Like the Hooded Oriole, they search the blossoms of ocotillo and desert agave for insects and nectar. They also visit chuparosa blossoms, inserting their bills down the tubular corolla in the manner of hummingbirds. Fruits of cactus, agave, and yucca are eaten and provide an important source of

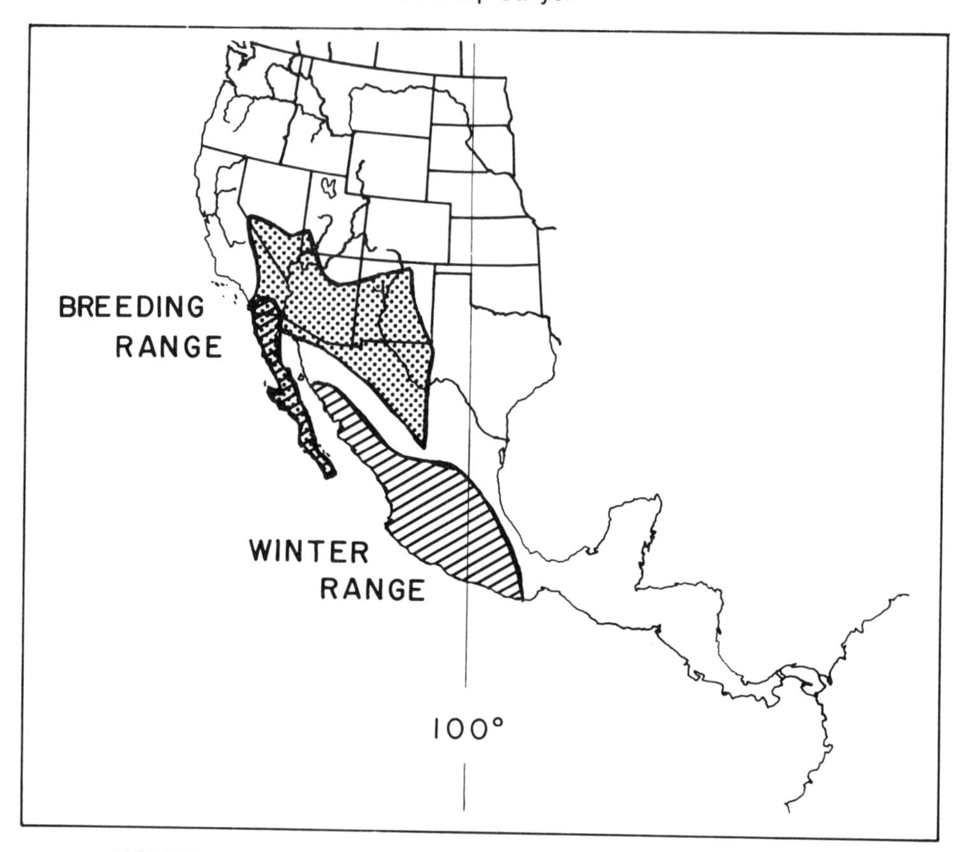

FIGURE 85 Map illustrating overlap of Scott's Oriole breeding and wintering ranges at Deep Canyon. A few Scott's Orioles winter at Deep Canyon, but most are summer visitors.

both water and energy. Orioles often forage in low desert scrub, searching for insects on or close to the ground. Because of the moisture contained in their diet, they can survive the hot, dry summer without drinking. Orioles do drink if water is available, however, and are often seen near streams in June.

Scrub Jays interfere with Scott's Orioles by robbing their nests and by driving the adults away from food. On 25 March 1978, Scott's Orioles in Deep Canyon's piñon-juniper habitat were visiting the pale blossoms of the desert apricot, which were aswarm with bees and other insects. Several times, Scrub Jays attacked and drove away the smaller orioles. Whether this was a manifestation of interspecific competition is uncertain. The jays were not feeding at the blossoms but may have been taking insects from the inner foliage. Jays and orioles are both omnivorous, take many of the same types of food, and thus might compete for food.

Scott's Orioles breed throughout the southwest in the arid woodlands that cover desert slopes and plateaus. In California, they nest principally in the tree yucca "forests" and the piñon-juniper association. At Deep Canyon, most nests are found in the piñon-juniper and lower plateau habitats, but a few pairs nest on the alluvial plain and in the chaparral.

Nests are shallow cups, typically built from stiff yucca fibers. They are slung in yucca beneath the spearlike leaves, in fan palms, or in the dried inflorescences of nolinas. No nests have been found in piñon trees at Deep Canyon, although elsewhere piñons are preferred (Grinnell and Miller 1944). Nest construction occurs between mid-April and late May. Early nesters may produce a second brood, whereas late nesters are probably single-brooded. Both sexes help in nest building, but the female alone incubates.

Fall migration begins in August or September, with most birds leaving Deep Canyon for the winter. A few remain behind in the piñon-juniper and lower plateau habitats.

Northern Oriole. *Icterus galbula*
Males 31–43 grams; females 27–37 grams

Synonyms: Bullock('s) Oriole; *Icterus bullocki(i).*

Range: Breeds in western North America from southern British Columbia, southern Alberta, and southwestern Saskatchewan south to northern Baja California and the southern part of the Mexican Plateau. Winters from southern Mexico through Central America to central Colombia and northwestern Venezuela.

Deep Canyon: Summer visitor and migrant.

Northern Orioles are much less numerous at Deep Canyon than either Scott's or Hooded Orioles. They occur regularly during spring migration in the desert washes, where they feed among the ocotillo blossoms and palo verde. Usually only one or two are seen during a morning walk through the desert wash habitat. For nesting, this species prefers riparian and oak woodland, especially where trees are large and well spaced. Water is frequently available near the nest site but is not essential. Sycamores, cottonwoods, willows, and deciduous oaks are specially favored nest sites, but live oaks, orchard trees, and occasionally conifers are used (Grinnell and Miller 1944).

Brewer's Blackbird. *Euphagus cyanocephalus*
Males 69–75 grams; females 52–69 grams

Synonyms: Brewer Blackbird; *Scolecophagus cyanocephalus.*

Range: Breeds from southwestern Canada east to western Ontario and south to southwestern United States and northwestern Baja California. Winters throughout breeding range, southeastern United States, and Mexico.

Deep Canyon: Permanent resident.

The species name, *cyanocephalus*, means "blue-headed" and is an allusion to the purplish gloss about the male's head. Brewer's Blackbirds are most often seen on lawns and golf courses, walking about in flocks. Birds sometimes wander into the sandy, desert regions adjacent to developed areas but do not stay long. They nest on the Deep Canyon Transect near water, usually around the developed areas. Brewer's Blackbirds frequently nest in small colonies, and males are sometimes polygamous.

Brown-headed Cowbird. *Molothrus ater*
Males 32–42 grams; females 27–37 grams

Synonyms: Cowbird; Dwarf Brown-headed Cowbird.

Range: From northeastern British Columbia and southern Mackenzie southeastward to southern Nova Scotia and south to Louisiana, northern Baja California, and central Mexico. Winters from central California eastward through southern Great Lakes region to Connecticut and south to southern Mexico.

Deep Canyon: Sparse resident.

Prior to 1915, cowbirds were rare in southern California. They were not even encountered during Grinnell and Swarth's (1913) survey of the Deep Canyon region. Subsequently, their numbers have increased phenomenally, especially in southwestern California. At Deep Canyon, they still are not numerous and are seldom found away from lawns and golf courses, where they associate with Brewer's Blackbirds. In spring, cowbirds disperse widely, and usually only single birds or pairs are seen at Deep Canyon during the breeding season.

Instead of building their own nests, Brown-headed Cowbirds deposit their eggs in the nests of other species, especially sparrows, warblers, and vireos. They are known to parasitize over 200 North American species and share this reproductive method (brood parasitism) with several other avian groups: the European cuckoos, some ducks, honeyguides, and a few Old World weaver finches. Each female cowbird lays a single egg in the host's nest. Accordingly, multiple cowbird eggs in a single nest are the work of more than one female.

How many eggs does a cowbird lay during a single breeding season? This question is difficult to answer for obvious reasons. The number should resemble that of other temperate passerines (8 to 12), and claims that females lay 30 eggs per season are based on circumstantial evidence. Payne (1965) histologically examined the ovaries of cowbirds. He concluded that several clutches are produced per season, with an interval between clutches of a few days to a few weeks. Average clutch size in Michigan was 3.1, and typical females produced 10 to 12 eggs per season.

Although I found no parasitized nests at Deep Canyon, the patrolling of males and females and their nuptial calls indicate that they breed there. Patrolling pairs were seen in the chaparral along Palm Canyon west of Deep Canyon, where Bell's Vireos formerly were abundant, and near Garnet Queen Creek.

TANAGERS—THRAUPIDAE

Only four of the more than 200 species in this large, tropical American family have been recorded from California. All are medium-sized, brightly colored birds with heavier, less tapered bills than orioles. Tanagers eat fruit, nectar, and insects. Members of this family are usually monogamous, and many remain in pairs all year.

Western Tanager. *Piranga ludovicianus*
28–32 grams

Synonym: Louisiana Tanager.

Range: Breeds from southern Alaska and central Mackenzie south to southwestern United States and northern Baja California. Winters from central Mexico south on the Pacific side of the Continental Divide in Central America to northwestern Costa Rica; casually north to California, Arizona, and Texas.

Deep Canyon: Migrant and summer visitor.

Adult male Western Tanagers are brightly colored, with red heads, yellow bodies, and black wings and tails (see colorplate). Their bright plumage makes them conspicuous and creates a false sense of abundance. They are uncommon on the Deep Canyon Transect. Western Tanagers occur throughout the transect during spring migration, from early through late May. Fall migration occurs chiefly from late August to mid-September.

Preferred nesting sites are located in fairly open coniferous forests or, less commonly, in piñon woodland. At Deep Canyon, nests have been found in the coniferous forest near Garnet Queen Creek. Usually, one or two pairs breed in the trees where the Santa Rosa Mountain road crosses the creek. Some individuals occur in the piñon-juniper woodland throughout the breeding season and may breed there. But, as yet, no nests have been found.

The nest, a bulky cup of twigs, rootlets, and moss, is often placed far out on the horizonal limb of a conifer, usually 3 to 10 m up. The clutch of three to five eggs is incubated by the female alone, but both parents feed the young.

GROSBEAKS, FINCHES, SPARROWS, AND BUNTINGS—FRINGILLIDAE

This large and diverse family of over 400 species occurs worldwide, except for Antarctica and a few remote oceanic islands. Most species are adapted for eating seeds and have heavy, conical bills with the cutting edges angled at the base. Fringillidae, the most recently evolved birds, trace their beginnings to the Miocene (25 to 30 million years ago), when seed-bearing plants suddenly came into prominence. Following the Miocene, grasses and sedges spread rapidly throughout the world and were followed by seed-eating birds.

Generalizations are seldom universal in such diverse groups. With a few exceptions, however, most species of this family share the following characteristics: they forage on or near the ground, are monogamous, establish territories during the breeding season, and build cup nests. Many temperate species are migratory and spend the winter in flocks.

The Fringillidae of Deep Canyon can be divided into three subfamilies. In the order listed in Appendix I, they are: 1) the *Cardinalinae*—the grosbeaks and the Lazuli Bunting; 2) the *Carduelinae* (cardueline finches)—Purple Finch through Red Crossbill; and 3) the *Fringillinae* or *Emberizinae* (fringillids)—towhees through Lincoln's Sparrow. Cardueline finches are adapted for dealing with hard seeds. They have powerful gizzards, strong skulls, stout conical bills (except the crossbills), and large jaw muscles. The bill of the Red Crossbill is adapted for extracting pine seeds from cones. Carduelines eat mainly seeds, even during the breeding season, and feed their nestlings on regurgitated seeds alone or on seeds mixed with insects. They nest solitarily or in loose colonies, with each pair defending only a small space around the nest. The fringillids are much less dependent upon seeds than are cardueline finches. Their diet changes from predominately seeds in winter to insects in summer. Their bills tend to be more slender than those of carduelines, an adaptation for feeding on insects. In many species, the mouth's roof has a bony hump against which seeds can be crushed, a substitute for the powerful cardueline gizzard. Fringillids feed their nestlings insects exclusively at first, later switching over to unripened seeds. They are strictly territorial when breeding but are gregarious at other seasons.

Black-headed Grosbeak. *Pheucticus melanocephalus*
35–40 grams

Synonyms: Pacific Black-headed Grosbeak; *Zamelodia melanocephala*; *Hedymeles melanocephalus*.

Range: Breeds from southwestern Canada and central Nebraska south to southern Mexico. Winters from northern Mexico and Louisiana south to southern Mexico.

Deep Canyon: Summer visitor and common migrant.

In spring, migrating Black-headed Grosbeaks pass through the Deep Canyon Transect between early April to mid-May. At this time, they are common to abundant in the desert washes, though less numerous elsewhere. The fall migration occurs in late August and September and is much less obvious than the spring migration.

Adult male grosbeaks are conspicuous because of their bright plumage and powerful song. Males sometimes sing in flight, and both sexes are reported to sing on the nest. At Deep Canyon, Black-headed Grosbeaks nest along the streams that flow from the coniferous forest. Nest construction, by the female alone, probably begins in late April or early May, judging from the advent of juveniles in late June. On 26 June, young were following adults through the riparian vegetation along Omstott Creek, loudly giving the begging note—*wheeoo.*

Lazuli Bunting. *Passerina amoena*
13–15 grams

Synonym: *Cyanospiza amoena.*

Range: Breeds from southern British Columbia across southern Canada to central North Dakota and south to northwestern Baja California, central Arizona, and central Texas. Winters from southern Baja California and southern Arizona south to Guerrero and central Veracruz.

Deep Canyon: Migrant and possible summer visitor.

Lazuli Buntings are seldom seen at Deep Canyon, although they may be expected at all transect elevations. For breeding, they prefer broken chaparral, other low hillside vegetation, and clumps of bushes in and about water courses. No nests have been found at Deep Canyon, but the presence of a singing male in the chaparral bordering Deep Canyon Creek on 28 June suggests local breeding.

Purple Finch. *Carpodacus purpureus*
20–23 grams

Synonym: California Purple Finch.

Range: Breeds from northern British Columbia eastward to Newfoundland and south on Pacific coast to northern Baja California; in east extends south to Great Lakes area and Maryland. Winters from southern Canada south through United States.

Deep Canyon: Summer visitor in coniferous forest; winter visitor in piñon-juniper woodland.

During summer, all three species of *Carpodacus* finches occur in the Santa Rosa Mountains' coniferous forest. Separating them in the field is sometimes next to impossible (especially the female and immature Cassin's and Purple Finches), but the following hints will help. The House Finch's bill is stout (almost as deep as long), whereas those of both Cassin's and Purple Finch are longer and thinner. The House Finch's tail is square-cut or slightly rounded, whereas those of the Cassin's and Purple Finch are notched. Male House and Cassin's Finches have streaked breasts, in contrast with the Purple Finch's clear breast. Undertail coverts of Cassin's Finch are streaked, and those of Purple Finch are clear (an excellent mark, if it can be seen). In flight the call of the male Purple Finch is a single *prit,* that of Cassin's Finch, a whistled *chee-wheet.*

The breeding population of Deep Canyon's Purple Finches consists of a few pairs that nest in the coniferous forest. During the winter, Purple Finches leave the high mountains and appear singly or in small flocks in the piñon-juniper habitat.

Cassin's Finch. *Carpodacus cassinii*
24–28 grams

Synonym: Cassin Purple Finch.

Range: Breeds in interior mountains of western North America from southern British Columbia and southwestern Alberta south to northern Baja California, southern Nevada, northern Arizona, central and northern New Mexico. Winters at lower elevations throughout breeding range and south through Mexican highlands.

Deep Canyon: Summer visitor and probable migrant.

This is the most numerous *Carpodacus* finch in Deep Canyon's coniferous forest. From April until mid-July, Cassin's Finches are common in the open forest in and about meadows, but usually away from streams. On 18 June 1979, young just out of the nest were following adults at Stump Spring Meadow. By late July, most of the Cassin's Finches leave the pines, which suggests that most pairs produce a single brood. A few continue to be seen until late October. Buds, especially the needle buds of conifers, are the Cassin's Finch's preferred food. But this species also forages on dry ground or in meadow grass.

House Finch. *Carpodacus mexicanus*
18–21 grams

Synonyms: Common House Finch; Linnet.

Range: Resident from southwestern British Columbia, central Idaho, and northern Wyoming south to southern Baja California, Guerrero, and central Oaxaca.

Deep Canyon: Resident.

The House Finch is gregarious, noisy, and abundant around towns and cities. It shares these characteristics with the introduced House Sparrow and Starling. This has led to the widespread but erroneous belief that the House Finch is similarly not native—a case of "guilt by association."

This attractive and tame bird (see colorplate) was widespread throughout the western United States before the coming of industrial man. With the advent of towns and cities, the adaptable House Finch began nesting in man-made structures and doubtlessly increased its numbers. Today it is one of the most conspicuous of the avian city dwellers. It has not forsaken its traditional haunts, however, and still lives—as it did before the arrival of man—in deserts, where it nests in cactus and spinescent trees. At Deep Canyon, the House Finch is one of the most abundant birds. It occurs throughout the transect from the valley floor upslope to the jeffrey pine forest but is most numerous at the lower elevations (fig. 86).

House Finches occur in flocks throughout most of the year. They move about locally, making use of seasonal food and water supplies. Some

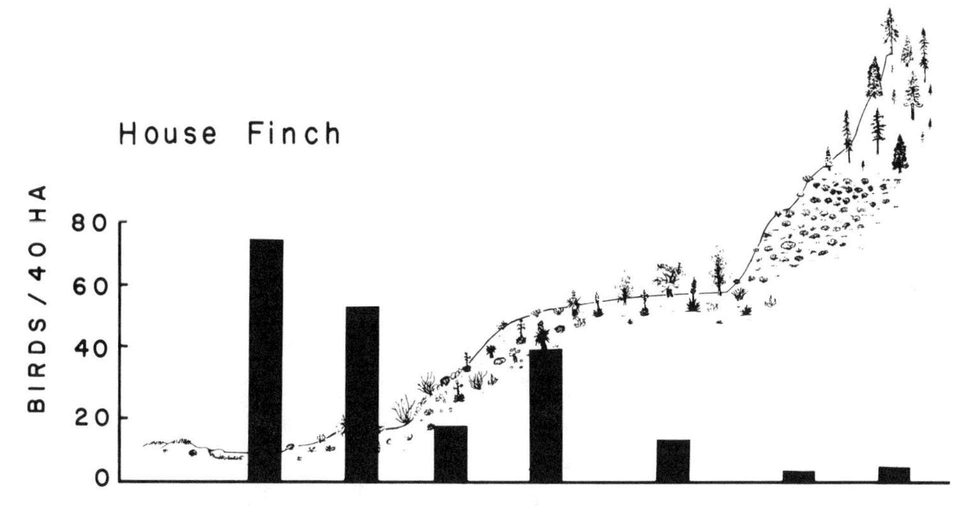

FIGURE 86 Mean annual density of House Finches at Deep Canyon. See fig. 38 for explanation of symbols.

appreciation of this seasonal wandering at Deep Canyon can be gained from table 44, which presents the seasonal density change throughout the transect. Many of the changes illustrated in table 44 (discussed in detail in the habitat chapters) result from the House Finch's daily need for water, either in the form of succulent food or as drinking water.

Like other cardinelines, House Finches are essentially ground-foraging seed-eaters. Seeds form the bulk of their diet at all times of the year, although succulent foods, especially green peppergrass (*Lepidium lasiocarpum*) and cactus fruits, are sometimes extensively eaten. During the breeding season, House Finches in Deep Canyon's dry desert washes are at least partially nectarivorous, obtaining nectar from chuparosa blossoms by pulling off the entire flower and "chewing" the base.

Bartholomew and Cade (1956) demonstrated experimentally that House Finches can survive without drinking water provided they are given succulent food. This ability contributes to their success in arid regions. Although House Finches are usually restricted to areas with surface water, Walsberg (1975) found them breeding in the Colorado Desert 7.4 km from the nearest drinking water. In this case, they were limited to areas with desert mistletoe, which they regularly consumed, apparently for its water content, as the birds left the area when the mistletoe stopped fruiting.

Near the Boyd Research Center, courtship begins in mid-February (±2 weeks), and there are usually nests with eggs or young by the middle of March. House Finch courtship involves song, special displays, posturing, and elaborate song flights over the nesting area. The male's courtship song, a varied, flowing, warbled series of whistles, is heard continuously

TABLE 44 Seasonal Variation in House Finch Density at Deep Canyon

Habitat	Birds/40 ha			
	Winter	Spring	Summer	Fall
Valley floor	215.0	4.4	103.6	14.7
Alluvial plain[a]	32.8	136.8	6.0	3.7
Rocky slopes	15.8	18.8	14.9	32.6
Lower plateau	1.5	154.8	13.4	10.5
Piñon-juniper	5.5	10.4	23.0	15.2
Chaparral	0	1.1	13.9	0.4
Coniferous forest	0	0	2.7	0

[a]Average of scrubland and desert wash habitats.

from first light to dusk. It is sung in flight or from a perch as the male follows the female. House Finch songs and song flights are among the most conspicuous spring events at Deep Canyon.

The House Finch's nest is a cup of fine weed and grass stems lined with feathers, string, wool, and/or plant down (see fig. 35). The nest, built by the female alone, is attractive when fresh, but since the young's droppings are not removed after the first three days, it gradually becomes quite soiled.

Smoke trees and jumping cholla are the preferred nest sites on Deep Canyon's alluvial plain. Where these plants form dense stands, House Finches nest in loose colonies with adjacent nests frequently only 3 m apart. House Finches seem to prefer smoke trees and cholla for the protection that these spiny plants offer from predators. The protection is not absolute, however, as snakes are able to climb cholla, and Cactus Wrens are at home among the cactus spines. Consequently, House Finch nestling mortality is generally high. To some extent, this is offset by the House Finch's high reproductive potential—three clutches per year sometimes being produced.

Clutch size is usually four (sometimes two to six), but occasionally one egg is infertile. Near the Boyd Research Center, House Finches are double-brooded and continue nesting into June. Higher up the transect, nesting apparently lasts longer, as I have seen House Finches copulating and carrying nesting material in the coniferous forest in early July.

Lesser Goldfinch. *Carduelis psaltria*
9–11 grams

Synonyms: Arkansas Goldfinch; Green-backed Goldfinch; *Astragalinus psaltria*; *Spinus psaltria.*

Range: Resident from southwestern Washington and western Oregon south and east to northwestern Oklahoma and central Texas, thence south through Mexico, Central America, and South America to northwestern Peru and northern Venezuela.

Deep Canyon: Common resident, but there is considerable local movement and vagrancy.

Lesser Goldfinches are found year-round at Deep Canyon. They are most abundant and widespread from mid-March through May, when they occur as spring migrants at the lower transect elevations (valley floor upslope through the lower plateau). Only a few Lesser Goldfinches have been seen between November to March, suggesting that most leave the area in winter.

The Lesser Goldfinch is the smallest of the four members of the genus *Carduelis.* It appears to be the most water-seeking of all the goldfinches, but this is probably a corollary of its occurrence in arid regions. Its nesting is limited to water sources. At Deep Canyon, it breeds in the cottonwoods along Deep Canyon Creek and in the golden-cup oaks along Garnet Queen Creek. In Arizona, its breeding season is prolonged, lasting from January through November (Phillips et al. 1964).

The Lesser Goldfinch typically forages near the ground, frequently feeding on the seed heads of low-growing composites. Miller and Stebbins (1964) saw only one individual feeding on buds and thought Lesser Goldfinches to be even more heavily dependent on seeds than House Finches. At Deep Canyon, scattered individuals are frequently seen feeding on the seed heads of brittlebush, but they also utilize succulent food. On 18 April 1980, I saw a flock of twenty-five goldfinches feeding on the ripening blossoms of pygmy cedar. In March of the same year, Lesser Goldfinches joined House Finches, orioles, and hummingbirds in feeding on chuparosa blossoms. In contrast to House Finches, goldfinches pierce the chuparosa blossom's base to extract the nectar, rather than tear off the entire flower.

Lawrence's Goldfinch. *Carduelis lawrencei*
9–12 grams

Synonyms: Lawrence Goldfinch; *Astragalinus lawrencei*; *Spinus lawrencei.*

Range: Breeds in California and northern Baja California, chiefly west of the Sierra Nevada and the southern deserts, and occasionally in western Arizona. Winters sporadically through the breeding range and eastward through central Arizona, New Mexico, and northern Sonora to western Texas.

Deep Canyon: Sparse summer visitor and winter visitor.

In California, Lawrence's Goldfinches are fairly common, but their numbers often vary from year to year in any given locality. Their distribution is notably discontinuous and their movements erratic (Grinnell and Miller 1944). At Deep Canyon, Lawrence's Goldfinches avoid the lower desert slopes of the Santa Rosa Mountains. They are seen most often as scattered individuals or small flocks, usually near water in the chaparral, as along Omstott Creek or Garnet Queen Creek, where they nest in the oak woodland.

Green-tailed Towhee. *Pipilo chlorura*
25–30 grams

Synonyms: *Oreospiza chlorura*; *Oberholseria chlorura*; *Chlorura chlorura*.

Range: Breeds in interior mountains from southeastern Washington and central Oregon eastward to central Wyoming and south to southern California. Winters from southern California, central Arizona, and western and southern Texas south to Baja California and Morelos, Mexico.

Deep Canyon: Migrant and summer visitor.

Migrating Green-tailed Towhees move through the transect during April and May. They are usually detected as single individuals on or near the ground or in fairly dense cover. This species generally prefers low chaparral-type brush, but during migration individuals are occasionally found in the dry desert washes.

Prior to this study, the San Jacinto Mountains were the southernmost known breeding location for this species. Mayhew (personal communication) found fledglings being fed by adults in chaparral at Garnet Queen Creek, indicating nearby breeding. Towhees also occur throughout the summer on Santa Rosa Mountain. The presence of territorial males in June near Stump Spring Meadow strongly suggests that some birds breed in the coniferous forest as well.

Rufous-sided Towhee. *Pipilo erythrophthalmus*
35–40 grams

Synonyms: Spurred Towhee; Spotted Towhee; San Diego Spotted Towhee; *Pipilo maculatus*.

Range: Breeds from southern British Columbia eastward across southern Canada and the northern United States to southwestern Maine, thence southward through United States, Baja California, and Mexico to Guatemala.

Deep Canyon: Permanent resident and probable winter visitor.

This red-eyed towhee seeks its food on the ground by scratching backward with both feet simultaneously in a peculiar hop. This manner of foraging, through layers of dead leaves and mulch, is a unique trait of the subfamily Fringillinae. It is most highly developed in members of the genus *Pipilo* and in the Fox Sparrow, yet it is seen occasionally in the Dark-eyed Junco and the White-crowned, Lincoln's, Vesper, and Song Sparrow (*Melospiza melodia*). The sound of steady scratching, as leaves are kicked out to the rear, is a clue to the presence of full-time scratchers like the Rufous-sided Towhee and the Fox Sparrow.

Because it requires leaf litter for foraging, the Rufous-sided Towhee is chiefly found in the chaparral and piñon-juniper habitats at Deep Canyon. Towhee density varies greatly with season in these two habitats. The change is reciprocal (table 45), suggesting that birds move between the two habitats. Some of the winter towhee influx in the piñon-juniper

TABLE 45 Seasonal Variation in Rufous-sided Towhee Density at Deep Canyon

Habitat	Birds/40 ha			
	Winter	Spring	Summer	Fall
Piñon-juniper	21.3	2.7	1.1	9.8
Chaparral	1.1	14.9	12.3	3.3

may represent arriving individuals of the northern race *P. erythrophthalmus montanus*, however. This race breeds in the Great Basin and winters as far south as Joshua Tree National Monument, across the Coachella Valley from Deep Canyon (Miller and Stebbins 1964). It prefers piñon-juniper habitat, whereas the resident Deep Canyon race (*P. erythrophthalmus megalonyx*) prefers chaparral.

Like that of most fringillids, the Rufous-sided Towhee's diet varies seasonally, from mainly plant material in winter, to a roughly fifty-fifty mixture of plant and animal matter in summer. Animal items include beetles, ants and other Hymenoptera, caterpillars, moths, grasshoppers, crickets, bugs, and flies (Martin et al. 1951). Plant material is mostly seeds, but fruits and berries are eaten in the fall.

The Rufous-sided Towhee is one of the most abundant breeding birds in Deep Canyon's chaparral habitat. The males perch conspicuously atop bushes or snags and sing, with monotonous persistence, a drawn-out, buzzy *chweeee*—suggesting the name towhee—or sometimes a *chup chup chup zeeeeeee*. Throughout the year, towhees utter a catlike call, *quee*, that is higher than the song. This frequently given call probably functions as a contact note and helps birds stay together as they move through the brushy thickets.

The towhee's nest, a stout cup of bark shreds, grasses, and rootlets, is placed on or near the ground in dense undergrowth. Typically, the nest is set in a small hollow made by the bird, with the rim at ground level. The female builds the nest, incubates the eggs, and broods the young with little or no assistance from the male.

Brown Towhee. *Pipilo fuscus*
40–48 grams

Synonyms: California Brown Towhee; Canyon Towhee; Anthony Towhee; *Pipilo crissalis*.

Range: Resident from southwestern Oregon south to southern Baja California and from western and central Arizona, northern New Mexico, southeastern Colorado, and central Texas south through Mexico to Colima and Oaxaca.

Deep Canyon: Permanent resident of the middle elevations.

This all-brown towhee requires dense, shrubby thickets for nesting, with adjacent open areas for foraging (Grinnell and Miller 1944). At Deep Canyon, accordingly, Brown Towhees are most frequently encountered in the chaparral, although they range downslope through the piñon-juniper to the lower plateau. They prefer the dense vegetation of canyon bottoms or streamsides bounded by open, desert slopes, such as along Omstott and Carrizo Creeks.

Their frequent association with riparian tracts suggests that Brown Towhees are dependent on water, although this is far from certain. Dawson (1954) thought that they required water daily, because captive birds were unable to survive twenty-four hours without water at high temperatures (39° C). Miller and Stebbins (1964) and Cord and Jehl (1979), however, never saw Brown Towhees drink, even in areas where water was available. At Deep Canyon, Brown Towhees occur far from water in the shrubby thickets of dry canyon bottoms and on the steep, chaparral-covered slopes. During spring and summer, they eat mostly insects and thus obtain considerable water in their food. On this diet, they can probably survive without drinking during mild weather, but they may require drinking water during prolonged hot spells.

Little information is available concerning the length of the breeding season at Deep Canyon. On 25 June, I saw a pair of adults followed closely by recently fledged young at Carrizo Creek. The next day, I saw another pair with young and a third pair feeding nestlings at Omstott Creek. This would place egg-laying around the first week of June. In Arizona, the Brown Towhee nests from March to September (Phillips et al. 1964), sometimes producing three broods. Brown Towhees are not only permanent residents wherever found but are remarkably sedentary. Pairs remain mated for life upon circumscribed territories (Marshall 1960). In contrast to the Rufous-sided Towhee, the Brown Towhee maintains a stable population throughout the year.

Abert's Towhee. *Pipilo aberti*
45–49 grams

Synonym: Abert Towhee.

Range: Resident from southeastern Nevada and Utah, central Arizona, and southwestern New Mexico south in the Colorado River drainage to southeastern California, northeastern Baja California, northwestern Sonora, and southeastern Arizona.

Deep Canyon: Sparse resident of low elevations, but local seasonal movements occur.

Abert's Towhees are found throughout the year at Deep Canyon, but they are rare, and the local birds shift about. Deep Canyon lies near the northwestern limit of this species' range in California (Whitewater, Riverside County) and thus is marginal habitat.

On 7 April 1980, I watched an Abert's Towhee gathering nesting material in Rubble Canyon Wash. It moved through the open, sandy part of the wash, running or hopping for about 2 to 3 m at a time, pausing for a few seconds, and then running again. After filling its beak with fine, dried grasses, it flew off to the east. This is the only indication that Abert's Towhees may nest on the Deep Canyon Transect.

The desert river valleys and bottomlands inhabited by Abert's Towhee are lower and hotter than the chaparral hillsides and canyons of the Brown Towhee. Accordingly, Abert's Towhees should be more heat-tolerant than their close relative. Dawson (1954) found that the physiological responses of Abert's Towhees and Brown Towhees to heat stress were similar, although Abert's Towhees were slightly more tolerant of high temperatures and drank significantly less water than Brown Towhees. Dawson's field observations revealed that on hot days both species become inactive and retreat to shade.

Although Abert's Towhee occurs in low-lying deserts, it is closely associated with riparian situations and is not a desert bird in the strictest sense. It appears to require water within its home range, and in desert scrub it often settles in areas bordered by riparian habitat (Grinnell and Miller 1944).

Black-throated Sparrow. *Amphispiza bilineata*
11–13 grams

Synonyms: Desert Sparrow; Desert Black-throated Sparrow.

Range: Breeds from southeastern Oregon, northern Utah, southwestern Wyoming, western and southern Colorado, and central Texas south to southern Baja California and through Mexico to Hidalgo. Winters from deserts of southern United States southward through breeding range.

Deep Canyon: Permanent resident, but some local seasonal movements.

Males and females of this small, attractive bird are alike in plumage (see colorplate). Immatures lack the adult's black bib and have a faded facial pattern and streaked breasts. Consequently, they somewhat resemble Sage Sparrows. First-year birds keep their juvenile plumage until October and, as they sometimes occur with fall flocks of Sage Sparrows, misidentification is possible.

Black-throated Sparrows are true desert birds, as their former name suggests. They occur in the hottest desert regions many miles from water, even in midsummer. What adaptations permit this tiny sparrow to be independent of water while larger species, such as the Gambel's Quail, are strictly limited by its availability? Although they seem remarkably well adapted to deserts, Black-throated Sparrows have no unusual thermoregulatory adjustments to heat (Weathers 1981). Their upper critical temperature (36° C) is no higher than that of nondesert birds. In the field, they show signs of heat stress at air temperatures above 38° C (100° F). Although they occasionally forage at high air temperatures (40° C), at such times they intersperse activity with repeated pauses in the shade and often gape their bills, pant, and hold their wings away from their bodies. In response to extreme heat, they cool themselves evaporatively, become less active, and retreat to shaded sites, including mammal burrows (Austin and Smith 1974).

Diet plays a major role in the Black-throated Sparrow's desert tolerance. Like other fringillids, it eats mostly seeds during fall and winter but lives on insects during summer. Whenever climatic conditions permit Black-throated Sparrows to find green vegetation or insects, they do not need to drink, even if air temperatures are high and solar radiation is intense (Smyth and Bartholomew 1966). When living on seeds, however, they seek water daily, even though air temperature is low.

Captive Black-throated Sparrows can survive on a diet of dried seeds without drinking water, provided air temperature is low. This ability, shared by more than a dozen small, seed-eating birds, is a by-product of small size rather than special physiological adjustments (see Brewer's Sparrow). For an explanation of how this and other desert adaptations work, see Bartholomew (1972), Dawson (1976), and Weathers (1981).

Throughout their range, Black-throated Sparrows prefer sparsely vegetated, strongly insolated terrain of exposed rock or gravel pavement. They avoid hot sinks and valley floors and confine their activities to the relatively cooler slopes and basins. Accordingly, though they are widespread at Deep Canyon, they are most abundant on the rocky slopes and lower plateau (fig. 87). Their density in both habitats changes seasonally, decreasing on the lower plateau in winter (see table 25) while increasing on the rocky slopes in spring (see table 21).

Like other fringillids, Black-throated Sparrows are gregarious during fall and winter, but territorial during spring and summer. In winter, they travel in small flocks, often accompanied by Sage, Brewer's, and White-crowned Sparrows. In February, pairs separate from the flocks

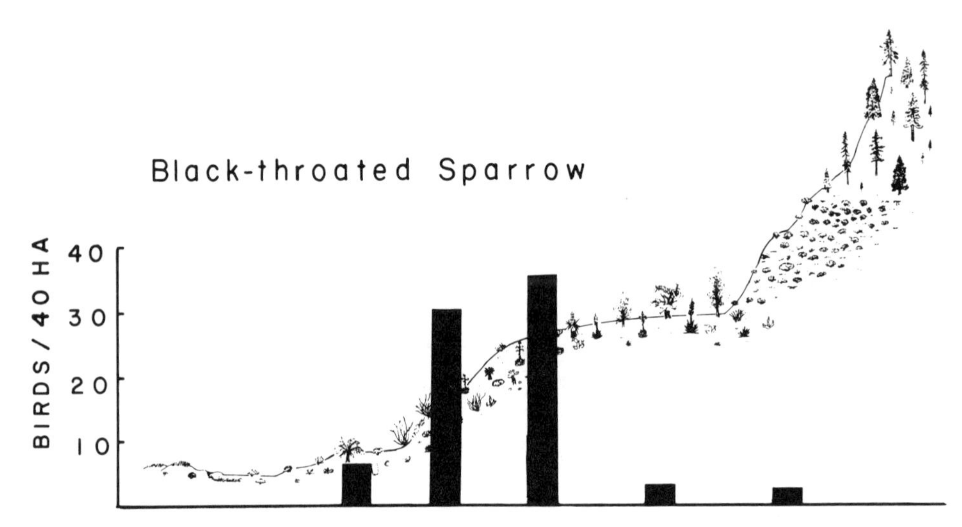

FIGURE 87 Mean annual density of Black-throated Sparrows at Deep Canyon. See fig. 38 for explanation of symbols.

and set up territories, and the males begin singing their persistent tinkling songs from atop low bushes. Through spring and summer, Black-throated Sparrows are strictly territorial and occur only in pairs or small family groups. By mid-June, males cease singing as the breeding season begins to draw to a close. Flocks form once again in late August and persist through winter until the following spring. I suspect some pairs remain together during winter.

Black-throated Sparrows nest from the alluvial plain up the Santa Rosa Mountains through the chaparral. Nests are typically located within about 0.25 m of the ground in a low, dense bush. They are neat, fairly deep cups made of light colored grasses and twigs and lined with plant down. Although built in dense bushes, nests are only partially shaded. On hot days, an adult usually shades the eggs or young by perching on the nest's rim.

Barbara Carlson (unpublished data) found that Black-throated Sparrows start nesting in February on Deep Canyon's rocky slopes. The percentage of successful nests varied from year to year as follows: 57 percent in 1978 (n = 10); 61 percent in 1979 (n = 13); and 22 percent in 1980 (n = 13). Because they are close to the ground, Black-throated Sparrow nests are subject to predation by antelope ground squirrels, leopard lizards (*Gambelia wislizenii*), and snakes. Cactus Wrens probably take a portion of the young as well. Females re-nest quickly following nest failures, and some birds may produce three to four clutches per season (compared to the usual two to three clutches). Nests with eggs have been found from 24 March through 16 June at Deep Canyon. Late nests are subject to the severest summer heat, and the survival of both adults and young must be tenuous at best.

Sage Sparrow. *Amphispiza belli*
16–20 grams

Synonyms: Bell Sparrow; California Bell Sparrow.

Range: Breeds from central interior Washington eastward to southwestern Wyoming and northwestern Colorado and southward to central Baja California, southern Nevada, northern Arizona, and northwestern New Mexico. Winters from central California, southern Great Basin, and southern Rocky Mountain regions south to central Baja California, northern Sonora, northeastern Chihuahua, and western Texas.

Deep Canyon: Occurs in three roles: permanent resident, migrant, and winter visitor.

This species' three inland races inhabit dry, fairly dense brushland. All forage on the ground, running between bushes with their black tails tilted upward. Rather than flying when approached, they often run to cover in a nearby bush.

The Sage Sparrow and its close relative, the Black-throated Sparrow, both inhabit waterless deserts during the breeding season. The Sage Sparrow tends to occupy regions that are slightly cooler and more shaded than those preferred by the Black-throated Sparrow. Although both species can survive on insects and succulent food without drinking water, the Black-throated Sparrow seems better adapted, physiologically, to aridity (cf. Smyth and Bartholomew 1966, Moldenhauer and Wiens 1970). The Black-throated Sparrow can survive on dry food without drinking water, while the Sage Sparrow cannot, and can tolerate 0.4 M NaCl, whereas the Sage Sparrow cannot drink water saltier than 0.20 M NaCl. Furthermore, the Black-throated Sparrow may be able to produce a more concentrated urine than the Sage Sparrow, although the two studies are not comparable in this respect. Because the Sage Sparrow lives in slightly cooler and more shaded sites, it may have less difficulty remaining in water balance. This is an attractive story. Unfortunately, it may not be true. There were important differences between the diet of Moldenhauer and Wien's (1970) Sage Sparrow study and Smyth and Bartholomew's (1966) Black-throated Sparrow study, which could account for the above results. Sage Sparrows were maintained on chick starter mash, whereas Black-throated Sparrows were given commercial mixed bird seed. Mash has a higher salt and nitrogen content than seeds and has been shown to require more water to process (Moldenhauer and Taylor 1973). Thus, if Sage Sparrows had been given seeds, they might have fared better.

Sage Sparrows occur on the Santa Rosa Mountains' chaparral-covered slopes throughout the year. But their density is low, and the breeding population contains only a few pairs. They are most abundant during fall migration when loose flocks of three to six can be found on the valley floor, foraging between the widely spaced creosote bushes. Fall migration occurs mainly during September and October, when the heat of summer has passed but succulent foods are nevertheless scarce. After fall migration, Sage Sparrow density on the valley floor declines markedly. Only a few birds remain through the winter. By March, most have departed for their breeding grounds to the north.

Dark-eyed Junco. *Junco hyemalis*
16–18 grams

Synonyms: Sierra Junco; Oregon Junco; *Junco oreganus.*

Range: Breeds throughout Canada south through the Appalachian Mountains in the east to northern Georgia (Slate-colored Junco) and in the west south to the mountains of northern Baja California (Oregon Junco). Winters from southeastern Alaska south throughout the United States to northern Mexico.

Deep Canyon: Occurs year-round, but in different roles: resident, summer visitor, migrant, and winter visitor.

Dark-eyed Juncos are abundant in the coniferous forests of the Sierra Nevada and southern California mountains. Their black heads and white outer tail feathers are familiar sights to all. They forage on the ground by gleaning seeds and insects, seldom scratching like other fringillids. They are rather tame. When flushed, they typically fly to a nearby tree or shrub.

Dark-eyed Juncos are found mainly above 1,000-m elevation at Deep Canyon. They are abundant in the jeffrey pine forest during spring and summer yet become scarce there in winter, when heavy snows push them downhill (see table 34). Dark-eyed Juncos occur in the piñon-juniper and chaparral habitats only as winter visitors. They begin arriving there in early September, spend the winter, and depart by mid-April. During their winter stay, juncos travel about in flocks, sometimes accompanied by White-crowned and Chipping Sparrows.

Juncos typically nest on the ground, building substantial cups of dried grasses and twigs in a small hollow dug by the bird. Nests are typically hidden among tree roots or bushes, frequently near stream-banks. On 20 June 1979, I found three junco nests (all with eggs) on 2 ha of open meadow near Stump Spring, Santa Rosa Mountain. Young emerge from late June to August, suggesting that some pairs rear two broods.

Chipping Sparrow. *Spizella passerina*
10–13 grams

Synonym: Western Chipping Sparrow.

Range: Breeds from central Yukon across Canada to Newfoundland south through the United States and mountains of Mexico and Central America to Nicaragua. Winters from about 36° north latitude southward.

Deep Canyon: Found year-round, but in different roles: winter visitor, migrant, and summer visitor.

Like Dark-eyed Juncos, Chipping Sparrows occur on the Deep Canyon Transect year-round but in differing seasonal roles. They arrive in the coniferous forest in late April, nest, and leave in September or October. On 18 June 1979, many downy young were following adults through the pines at Stump Spring Meadow, which would place egg laying in late May. A few Chipping Sparrows occur in the piñon-juniper woodland during summer and may nest there.

Chipping Sparrows occur on the Deep Canyon Transect as winter visitors, chiefly from September through April. During this period, they are most abundant in the piñon-juniper woodland. The few individuals wintering on the valley floor and alluvial plain might be confused with Brewer's Sparrows, which they closely resemble during winter. The Chipping Sparrow's rufous crown and black bill show only in spring and summer. The gray rump and narrow, black line through the eye are good field marks year-round.

Migrating Chipping Sparrows pass through the valley floor and alluvial plain of the Deep Canyon Transect between late March and May. They forage for seeds and insects on bare ground between creosote bushes, sometimes alone, sometimes in the company of other sparrows.

During fall migration, they avoid the parched lower elevations and travel south through the mountains. I found many migrants in the piñon-juniper habitat during September.

The jeffrey pine forest atop the Deep Canyon Transect is relatively dry, and the Chipping Sparrows that nest there frequently occur far from water. This suggests that they do not need to drink. Indeed, Chipping Sparrows are able to survive moderate temperatures on dried seeds alone (Moldenhauer and Taylor 1973, Dawson et al. 1979). During summer, they eat mainly insects and thus probably do not require water, although they drink when it is available.

Brewer's Sparrow. *Spizella breweri*
9–12 grams

Synonym: Brewer Sparrow.

Range: Breeds in higher basins and interior mountains of western North America from southwestern Yukon south to southern California, central Arizona, and northwestern New Mexico. Winters from southern California east to central Texas and south to southern Baja California, Jalisco, and Guanajuato.

Deep Canyon: Primarily a winter visitor and migrant.

Brewer's Sparrows begin arriving on the valley floor and alluvial plain in early September, when summer weather is still in force. These early winter visitors must contend with high air temperatures (>40° C) and limited water supplies. They forage beneath bushes and across open ground, gleaning seeds from among the dried remnants of annual vegetation. Only occasionally do they capture insects. Moist food is therefore limited at a time of increased demand for evaporative water loss. In captivity, Brewer's Sparrows can survive on dried seeds without drinking water, and they are more xerophilous than the congeneric Chipping Sparrow (Dawson et al. 1979). Whether they obtain enough insects in fall to meet their water needs is uncertain. Once the winter rains come and the annual plants germinate, Brewer's Sparrows begin feeding on the green seed-heads and become independent of drinking water.

In winter, Brewer's Sparrows travel in flocks of three to six (family groups?), often in the company of White-crowned and Black-throated Sparrows. They forage for insects in dense, desert shrubs more often than their companions.

These sparrows begin singing their buzzy, canarylike song before leaving Deep Canyon in spring, and their choruses ring forth across the alluvial plain throughout the day. By early May, all but a few birds have left, and those that remain are silent.

Brewer's Sparrows prefer extensive tracts of sagebrush (*Artemisia tridentata*) for breeding, but comparable low brushland will suffice. They breed west of the Deep Canyon Transect at Kenworthy but apparently shun the Santa Rosa Mountains' arid slopes, as no evidence of nesting has been found there.

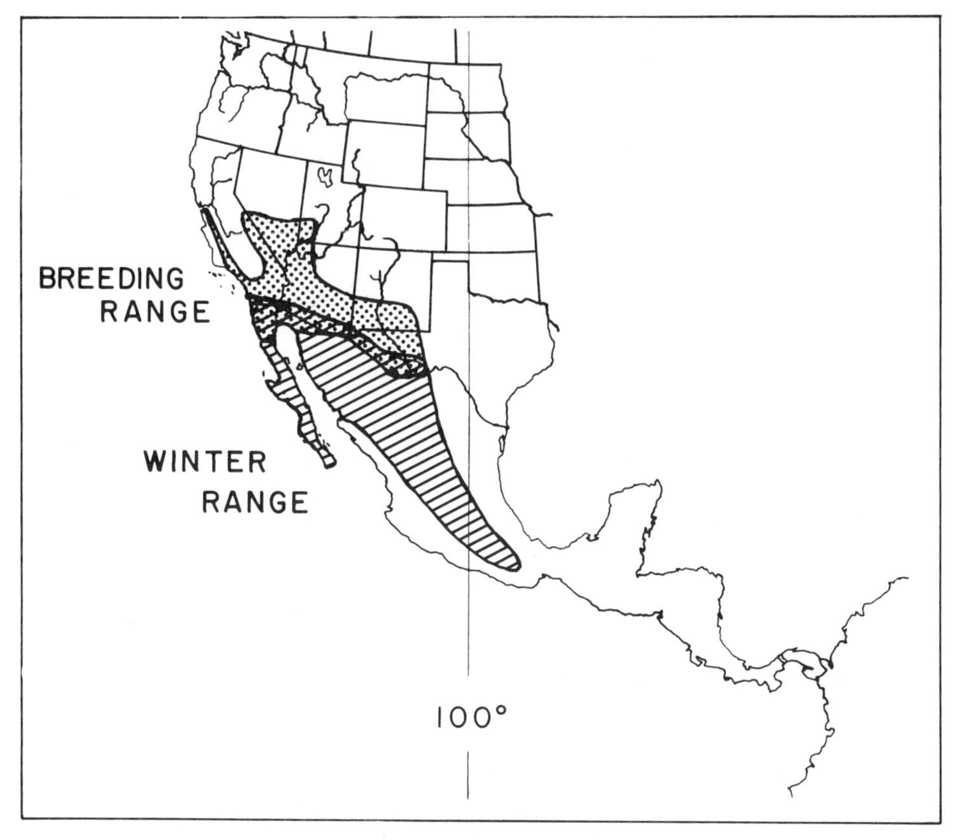

FIGURE 88 Map showing principal breeding and wintering ranges of the Black-chinned Sparrow.

Black-chinned Sparrow. *Spizella atrogularis*
10–12 grams

Synonym: California Black-chinned Sparrow.

Range: Breeds from central California, southern Nevada, central Arizona, and southern New Mexico south to northern Baja California, Guerrero, and Oaxaca. Winters from southern California, southern Arizona, and central Texas southward (fig. 88).

Deep Canyon: Sparse migrant, abundant summer visitor.

This diminutive sparrow arrives at Deep Canyon in April. From then until September, it is the most abundant bird of the steep, chaparral-covered slopes. Black-chinned Sparrows are especially abundant in the desert ceanothus along the Santa Rosa Mountain road, where the attractive males pour forth their song throughout the day from atop the taller shrubs.

Like other members of the subfamily Emberizinae, this sparrow eats chiefly insects during the spring and summer, gleaning them from the foliage and ground within the dense chaparral. A moist diet and a shaded environment permit Black-chinned Sparrows to survive without drinking water. Physiologically, their water economy should resemble that of their congener, the Brewer's Sparrow (see Dawson et al. 1979).

By all accounts, nests of the Black-chinned Sparrow are difficult to find. The species is secretive and usually stays out of view in the dense chaparral vegetation. The nests are loosely constructed cups placed low in a shrub. The usual clutch is two to four, but few other breeding details are available (Harrison 1978).

White-crowned Sparrow. *Zonotrichia leucophrys*
24–30 grams

Synonym: Gambel's Sparrow.

Range: Breeds from northern Alaska east along tree line to northern Labrador and south to southern California, southern Sierra Nevadas, and northern New Mexico; in east, south to central Manitoba and southern Quebec. Winters from southern British Columbia, southwestern Washington, southern Idaho, Wyoming, Kansas, Missouri, Kentucky, and western North Carolina south to southern Baja California, Jalisco, Michoacan, Querétaro, the Gulf Coast of the United States, and Cuba (fig. 89).

Deep Canyon: Widespread winter visitor and migrant.

Two races of the White-crowned Sparrow occur at Deep Canyon. The most abundant race, *Z. leucophrys gambelii* (fig. 90) breeds far to the north and, of the two races, is the more common winter visitor throughout the transect. The other race, the Mountain White-crowned Sparrow (*Z. leucophrys oriantha*), breeds in the Sierra Nevada Mountains and passes through the transect in low numbers during spring migration.

White-crowned Sparrows have been detected at Deep Canyon as early as 25 September and as late as 7 May. Abundant during their period of winter residence, they settle in dense vegetation (such as is found around homes), form cohesive flocks, and sing throughout the day. These traits make them Deep Canyon's most conspicuous winter visitor.

Wintering White-crowned Sparrows prefer low, scattered, brushy cover. They forage in flocks over open ground, often accompanied by Black-throated Sparrows. While foraging, they stay close to bushes, to which they retreat in contagious alarm when danger threatens. This species avoids forests and barren deserts and settles mainly in Deep Canyon's valley floor, alluvial plain, piñon-juniper, and chaparral habitats (fig. 91). Moreover, the weather affects the sparrows such that the dates of primary winter residence vary with elevation (fig. 92). During fall, temperatures on the valley floor are still high, and arriving white-crowns settle first at the higher, cooler elevations. In January, as the upper mountain slopes become cold, white-crowns leave the piñon-juniper and chaparral habitats for the warmer low desert. A large increase in white-crowns is seen at the low elevations in March and April (fig. 92), as migrants move through the Coachella Valley.

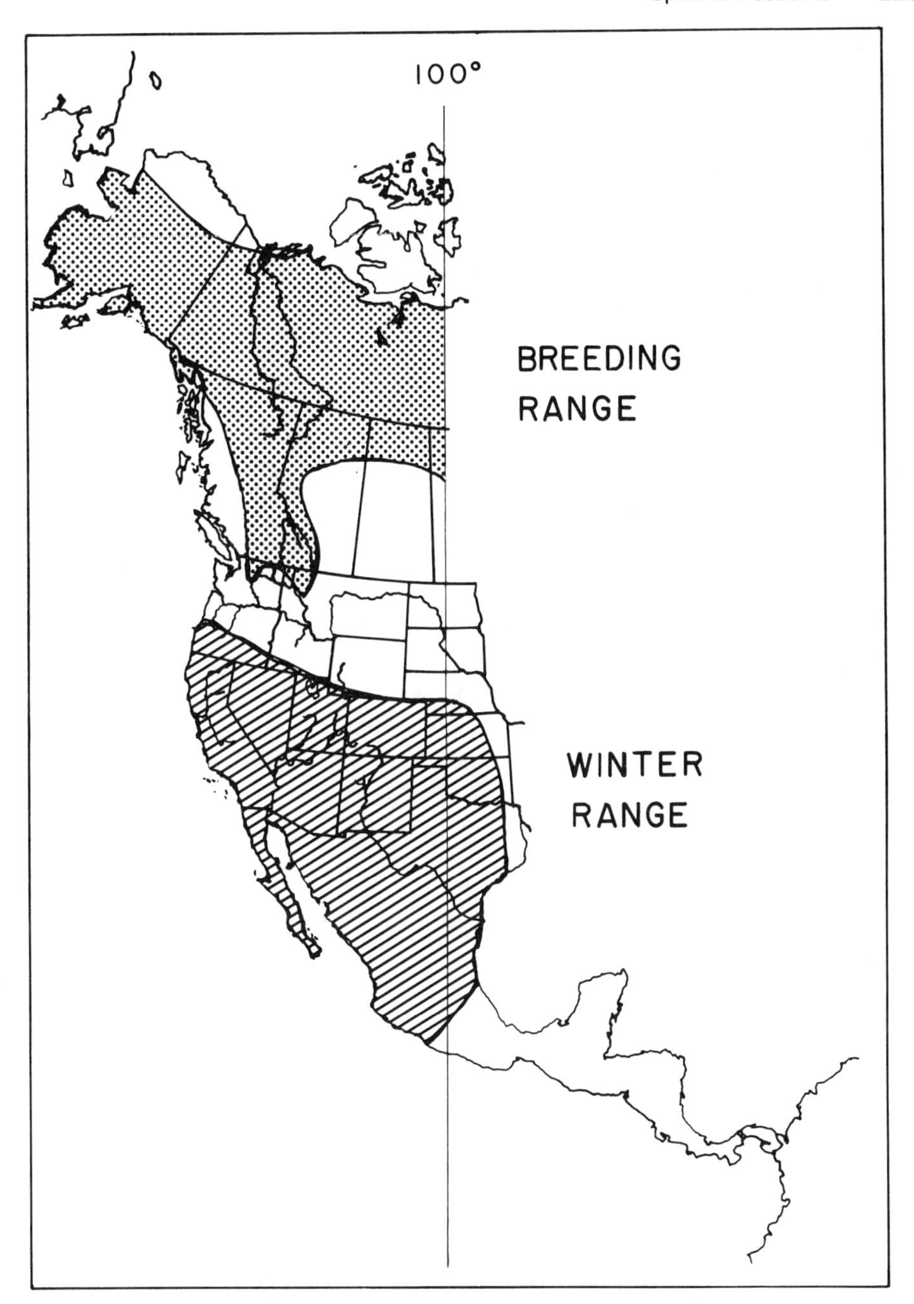

FIGURE 89 Map showing principal breeding and wintering ranges of the White-crowned Sparrow (*Zonotrichia leucophrys gambelii*), an abundant winter visitor at Deep Canyon.

White-crowned Sparrows sing throughout winter and especially vigorously at roost time when they retreat to dense trees and shrubs. Singing while going to roost helps attract several flocks to the same site,

FIGURE 90 White-crowned Sparrows prefer dense cover for roosting.

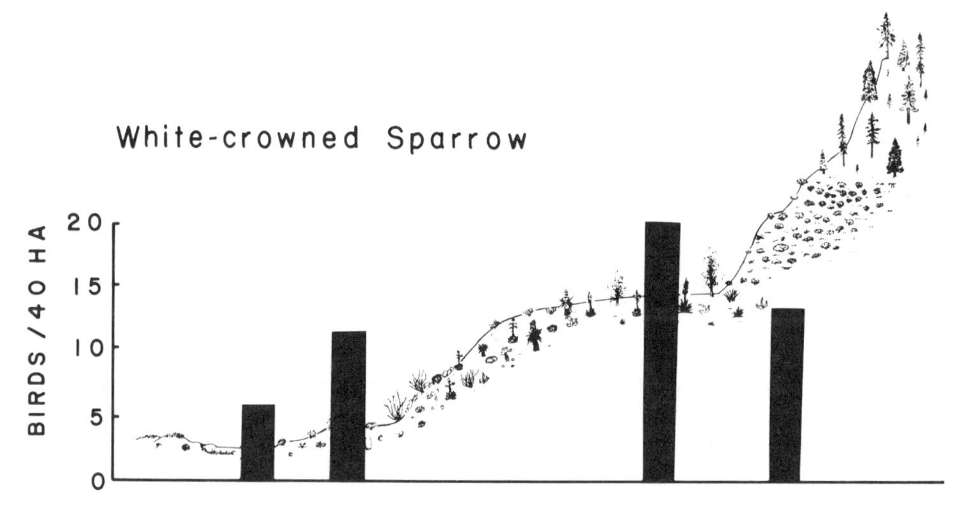

FIGURE 91 Mean density of White-crowned Sparrows at Deep Canyon during fall and winter. See fig. 38 for explanation of symbols.

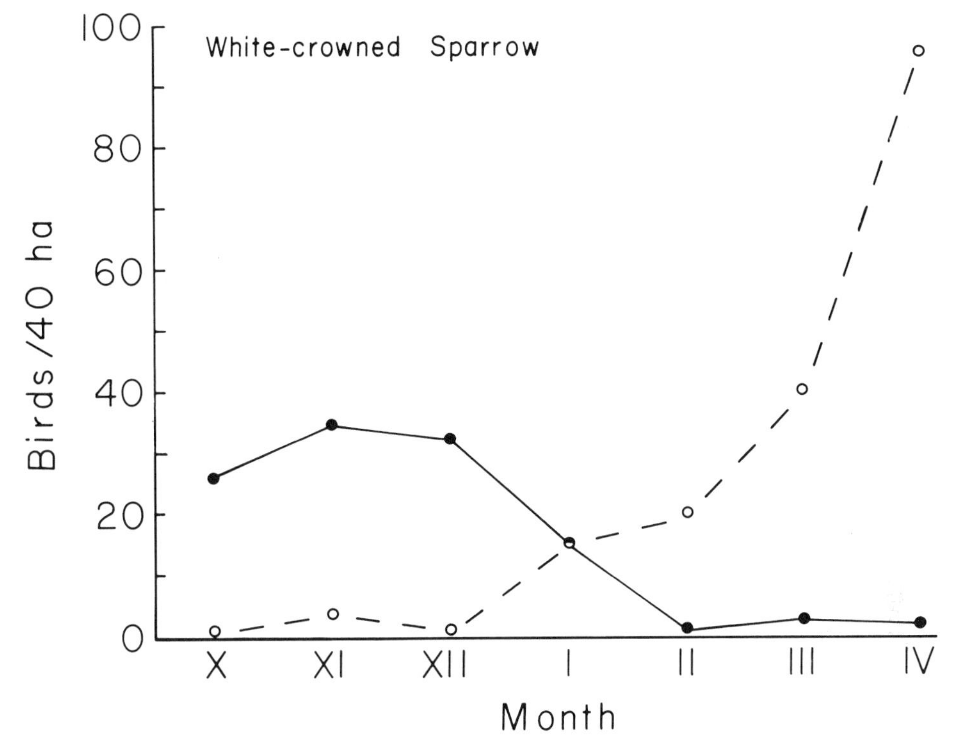

FIGURE 92 Relation of White-crowned Sparrow density at Deep Canyon to time of year. Dots and solid line are combined data for chaparral and piñon-juniper habitats. Circles and dashed line are combined data for alluvial plain and valley floor habitats.

and I have found over 100 sparrows roosting in a single palo verde tree. As the light begins to fade and the roost becomes quiet, the sparrows respond with sharp *chip* notes if disturbed.

Fox Sparrow. *Passerella iliaca*
28–33 grams

Synonyms: A different name for each of the eighteen races.

Range: Breeds from arctic tree line south; in west from northwestern Alaska south through mountains to central Colorado and southern California, in east to central Ontario and southern Quebec. Winters from southern British Columbia, southern Utah, Colorado, the lower Missouri Valley, the southern Great Lakes Basin, and southern New Brunswick south to northern Baja California, southern Arizona, southern Texas, the Gulf Coast, and central Florida.

Deep Canyon: Primarily a winter visitor, but also a summer visitor in the coniferous forest.

FIGURE 93 Fox Sparrows forage by using their long hind claw to scratch through leaf litter.

This large, dark sparrow is nearly the size of the Rufous-sided Towhee. The sexes are alike in plumage, with uniformly dark backs and breasts spotted heavily with inverted V's gathered together into a central spot (fig. 93).

These sparrows pass the winter on the chaparral-covered slopes of the Santa Rosa Mountains. From early October until mid-April, they scratch vigorously through the leaf litter—aided by a long hind claw—searching for seeds and insects. When disturbed, Fox Sparrows plunge headlong into dense cover, giving a metallic *chick* as they fly. They respond strongly to "squeaking," with a dozen or more sometimes popping abruptly into view.

Fox Sparrows and Green-tail Towhees occur in similar habitats. In 1908, they were confined as breeding species to the San Jacinto Mountains and did not occur in the Santa Rosa Mountains. Grinnell and Swarth (1913) reported that Fox Sparrows were abundant in San Jacinto's Tahquitz and Round Valleys, where they were closely associated with bush chinquapin (*Castanopsis sempervirens*). They further noted, "In the higher parts of the Santa Rosa Mountains, where the fox sparrows might reasonably be expected to occur, there was no chinquapin, and none of the birds." Like the Green-tailed Towhee, the Fox Sparrow may have expanded its breeding range into the Santa Rosa Mountains since 1908. I found a few individuals of both species near Stump Spring Meadow from May through July. They were associated with thickets of snowbush (*Symphoricarpus parshii*), which somewhat resembles chinquapin. The towhees were obviously nesting there, but I am uncertain about the Fox Sparrows.

appendix one

SPECIES LIST

The birds observed on the Deep Canyon Transect through 1 July 1981 are listed in this appendix by habitat and month of occurrence. It provides a quick means of checking a particular species' status but has certain limits. For example, the Western Bluebird was found in all the major habitats and at all times of the year, but this does not mean that it occurs in every habitat throughout the year. Western Bluebirds are winter visitors in the lower elevations of the transect but year-round residents in the coniferous forest. Thus, although the list does indicate in which habitats or months birds have not been seen, you may need to refer to the species accounts or habitat chapters to determine the seasonal status of a bird in a particular habitat.

Appendix symbols are as follows: dots denote single occurrences, solid lines indicate multiple sightings in a given month, and dashed lines (given for terrestrial species only) indicate when a species is to be expected. I have not subdivided months. Consequently, even if a species was seen only during the last week of June, the entire month is still barred. The letter **N** denotes observation of one of the following indicators of breeding: an active nest; adults feeding downy, recently fledged young; or territorial birds carrying nesting material or food. The symbol ?**N** indicates that the species may nest on the transect. Body masses, used in energy calculations, are means derived from the literature and specimens in the Museum of Vertebrate Zoology, University of California, Berkeley.

Vernacular and scientific names follow the fifth edition of the Check-list of North American Birds (American Ornithologists' Union 1957) as revised by the 32d and 33d supplements (The Auk 90:411–419 [1973], and The Auk 93:875–879 [1976], respectively).

This list was compiled from observations made by many people over a period of roughly twenty years. The contributors are far too numerous to acknowledge individually, but a few warrant special mention. Chief among these is Wilbur W. Mayhew, professor of Biology at University of California, Riverside, and Director of the Philip L. Boyd Deep Canyon

Desert Research Center. Over the years, Professor Mayhew has spent countless hours walking hundreds of miles of transect lines at Deep Canyon and has contributed many of the observations reported here. Jan Zabriskie spent six years at the Research Center as the Resident Researcher and made many valuable observations. Other major contributors include Barbara Carlson, Toni Gull, Sue Fuller, Karen Sausman, Colin Wainwright, Andy Sanders, William O. Wirtz, and my wife, Debra Weathers.

Note added in proof: After this book was in proof, the 34th supplement to the American Ornithologists' Union Check-list of North American Birds appeared (The Auk 99:1cc-16cc [1982]). The 34th supplement presages the appearance of the 6th edition of the A.O.U. Check-list of North American Birds. The new A.O.U. checklist contains numerous taxonomic changes, especially among the order *Passeriformes*, and will supercede the classification scheme followed in this book.

SPECIES	Valley Floor	Human Habitats	Alluvial Plain	Rocky Slopes	Lower Plateau	Piñon-Juniper	Chaparral	Coniferous Forest	Streamside	J	F	M	A	M	J	J	A	S	O	N	D
GAVIIDAE																					
Common Loon (*Gavia immer*)		X												•							
Arctic Loon (*Gavia arctica*)		X																			
Red-throated Loon (*Gavia stellata*)		X																	•		
PODICIPEDIDAE																					
Horned Grebe (*Podiceps auritus*)		X																		•	
Eared Grebe (*Podiceps nigricollis*)		X																			
Western Grebe (*Aechmophorus occidentalis*)		X																			
Pied-billed Grebe (*Podilymbus podiceps*)		X																			
PELECANIDAE																					
White Pelican (*Pelecanus erythrorhynchos*)[a]	X	X																			
ARDEIDAE																					
Great Blue Heron (*Ardea herodias*)		X	X																	•	
Northern Green Heron (*Butorides striatus*)		X																			
Cattle Egret (*Bubulcus ibis*)		X										•									
Great Egret (*Casmerodius albus*)		X	X																		

[a]Migrating overhead.

SPECIES	Valley Floor	Human Habitats	Alluvial Plain	Rocky Slopes	Lower Plateau	Piñon-Juniper	Chaparral	Coniferous Forest	Streamside	J	F	M	A	M	J	J	A	S	O	N	D
Snowy Egret (*Egretta thula*)		X	X																		
American Bittern (*Botaurus lentiginosus*)		X														•					
ANATIDAE																					
Canada Goose (*Branta canadensis*)	X	X	X																		
Snow Goose (*Chen caerulescens*)		X																			
N Mallard (*Anas platyrhynchos*)		X																			
Pintail (*Anas acuta*)	X	X	X																		
American Green-winged Teal (*Anas crecca*)	X	X	X																		
Blue-winged Teal (*Anas discors*)		X										•									
Cinnamon Teal (*Anas cyanoptera*)	X	X																			
American Wigeon (*Anas americana*)		X																			
Northern Shoveler (*Anas clypeata*)		X																			
Wood Duck (*Aix sponsa*)		X																			
Redhead (*Aythya americana*)		X																			
Ring-necked Duck (*Aythya collaris*)		X																			
Lesser Scaup (*Aythya affinis*)		X																			
Bufflehead (*Bucephala albeola*)		X																			

SPECIES	Body Mass (g)	Valley Floor	Human Habitats	Alluvial Plain	Rocky Slopes	Lower Plateau	Piñon-Juniper	Chaparral	Coniferous Forest	Streamside	J	F	M	A	M	J	J	A	S	O	N	D
Ruddy Duck (*Oxyura jamaicensis*)	–		X														•					
CATHARTIDAE																						
Turkey Vulture (*Cathartes aura*)	1580	X	X	X	X	X	X	X														
ACCIPITRIDAE																						
Goshawk (*Accipiter gentilis*)	925								X							•						
Sharp-shinned Hawk (*Accipiter striatus*)	138		X	X	X	X				X												
N Cooper's Hawk (*Accipiter cooperii*)	350	X	X	X	X	X	X	X	X	X												
N Red-tailed Hawk (*Buteo jamaicensis*)	1063	X	X	X	X	X	X	X	X	X												
Swainson's Hawk (*Buteo swainsoni*)	861	X						X						•			•					
N Zone-tailed Hawk (*Buteo albonotatus*)	757							X	X	X												
Ferruginous Hawk (*Buteo regalis*)	1124	X																				•
N Golden Eagle (*Aquila chrysaetos*)	3605	X		X	X	X		X	X													
Bald Eagle (*Haliaeetus leucocephalus*)	4620	X											•									
Northern Harrier (*Circus cyaneus*)	389	X	X	X																		
Osprey (*Pandion haliaetus*)	1532		X																	•		

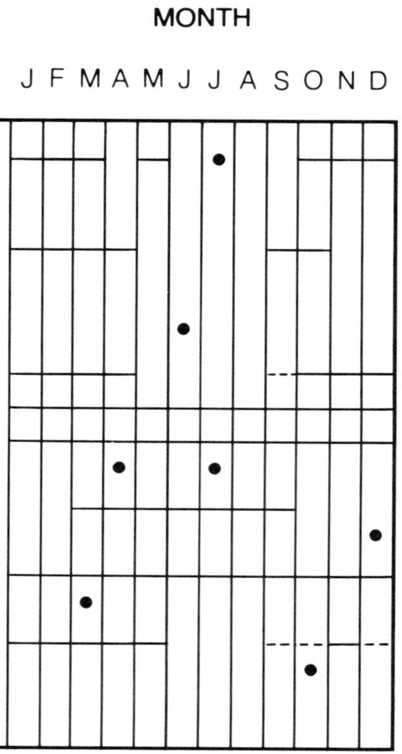

SPECIES	Body Mass (g)	Valley Floor	Human Habitats	Alluvial Plain	Rocky Slopes	Lower Plateau	Piñon-Juniper	Chaparral	Coniferous Fores	Streamside	J	F	M	A	M	J	J	A	S	O	N	D
FALCONIDAE																						
N Prairie Falcon (*Falco mexicanus*)	702	X	X	X	X	X	X			X												
Peregrine Falcon (*Falco peregrinus*)	815	X										●										
Merlin (*Falco columbarius*)	182				X					X												
N American Kestrel (*Falco sparverius*)	119	X	X	X	X	X	X	X	X	X												
PHASIANIDAE																						
N California Quail (*Lophortyx californicus*)	177			X		X	X	X		X												
N Gambel's Quail (*Lophortyx gambelii*)	171	X	X	X	X	X	X	X		X												
N Mountain Quail (*Oreortyx pictus*)	226					X	X	X	X	X												
RALLIDAE																						
American Coot (*Fulica americana*)	–		X																			
CHARADRIIDAE																						
N Killdeer (*Charadrius vociferus*)	85	X	X	X																		
SCOLOPACIDAE																						
Common Snipe (*Capella gallinago*)	–		X																			

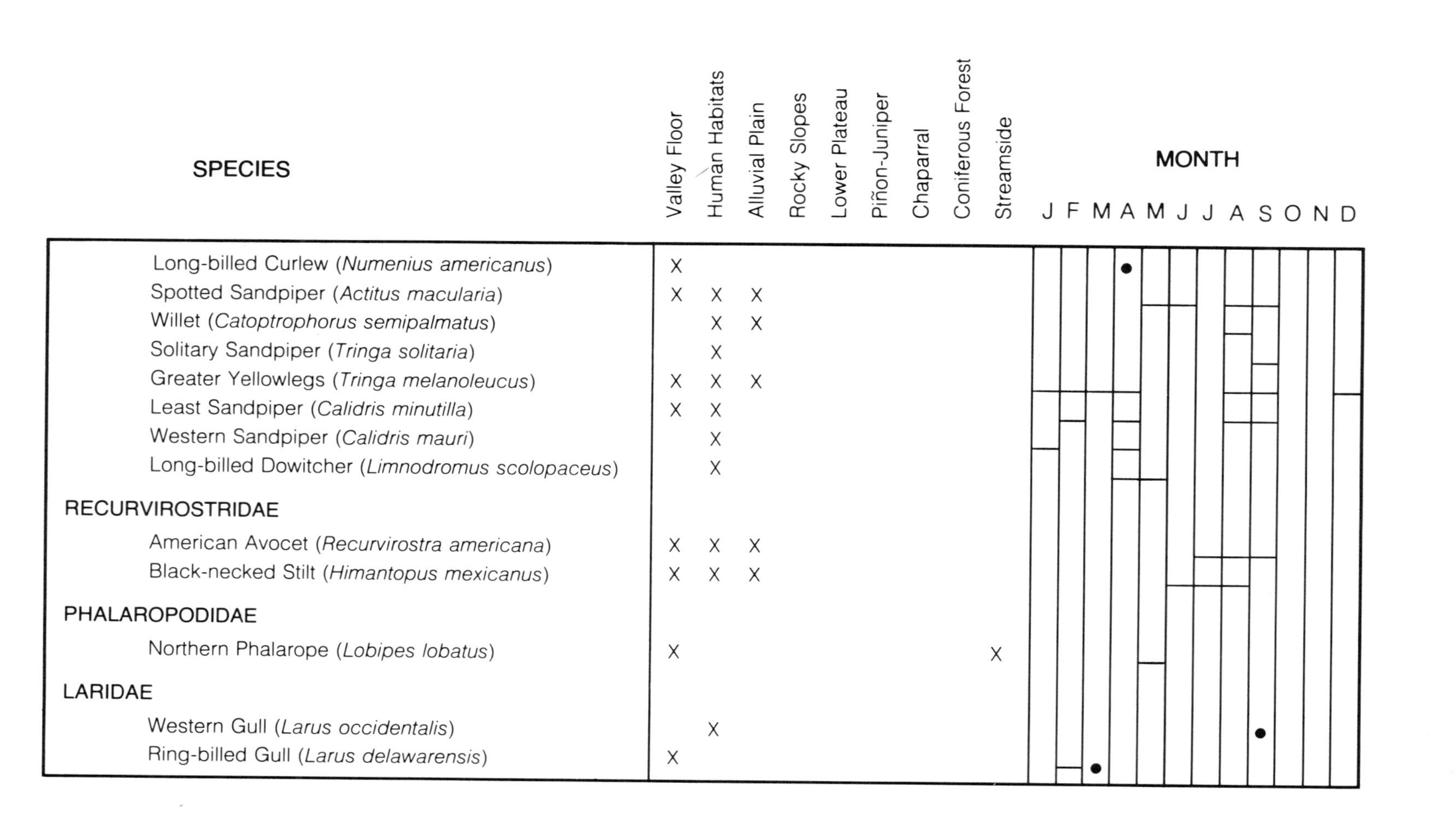

SPECIES	Valley Floor	Human Habitats	Alluvial Plain	Rocky Slopes	Lower Plateau	Piñon-Juniper	Chaparral	Coniferous Forest	Streamside
Long-billed Curlew (*Numenius americanus*)	X								
Spotted Sandpiper (*Actitus macularia*)	X	X	X						
Willet (*Catoptrophorus semipalmatus*)		X	X						
Solitary Sandpiper (*Tringa solitaria*)		X							
Greater Yellowlegs (*Tringa melanoleucus*)	X	X	X						
Least Sandpiper (*Calidris minutilla*)	X	X							
Western Sandpiper (*Calidris mauri*)		X							
Long-billed Dowitcher (*Limnodromus scolopaceus*)		X							
RECURVIROSTRIDAE									
American Avocet (*Recurvirostra americana*)	X	X	X						
Black-necked Stilt (*Himantopus mexicanus*)	X	X	X						
PHALAROPODIDAE									
Northern Phalarope (*Lobipes lobatus*)	X								X
LARIDAE									
Western Gull (*Larus occidentalis*)		X							
Ring-billed Gull (*Larus delawarensis*)	X								

SPECIES	Body Mass (g)	Valley Floor	Human Habitats	Alluvial Plain	Rocky Slopes	Lower Plateau	Piñon-Juniper	Chaparral	Coniferous Forest	Streamside	MONTH J F M A M J J A S O N D
Bonaparte's Gull (*Larus philadelphia*)	–		X								
Forster's Tern (*Sterna forsteri*)	–		X								
Black Tern (*Chlidonias niger*)	–		X								
COLUMBIDAE											
N Band-tailed Pigeon (*Columba fasciata*)	388							X	X	X	
N Rock Dove (*Columba livia*)	450	X	X	X		X	X			X	
N White-winged Dove (*Zenaida asiatica*)	154	X	X	X	X	X				X	
N Mourning Dove (*Zenaida macroura*)	120	X	X	X	X	X	X	X	X	X	
N Spotted Dove (*Streptopelia chinensis*)	140		X								
N Ground Dove (*Columbina passerina*)	36	X	X	X						X	
CUCULIDAE											
N Roadrunner (*Geococcyx californianus*)	300	X	X	X	X	X	X	X		X	
TYTONIDAE											
N Barn Owl (*Tyto alba*)	456	X	X	X			X				

SPECIES	Body Mass (g)	Valley Floor	Human Habitats	Alluvial Plain	Rocky Slopes	Lower Plateau	Piñon-Juniper	Chaparral	Coniferous Forest	Streamside
STRIGIDAE										
N Screech Owl (*Otus asio*)	148	X	X	X			X		X	
N Great Horned Owl (*Bubo virginianus*)	1026	X		X	X	X	X			
Elf Owl (*Micrathene whitneyi*)	43		X							
N Burrowing Owl (*Athene cunicularia*)	152	X		X						
Long-eared Owl (*Asio otus*)	262			X		X				
Short-eared Owl (*Asio flammeus*)	347	X		X						
CAPRIMULGIDAE										
N Poor-will (*Phalaenoptilus nuttallii*)	45	X	X	X	X	X	X	X		X
Common Nighthawk (*Chordeiles minor*)	69	X		X						
N Lesser Nighthawk (*Chordeiles acutipennis*)	48	X	X	X		X				X
APODIDAE										
Vaux's Swift (*Chaetura vauxi*)	18	X	X	X	X	X				X
N White-throated Swift (*Aeronautes saxatalis*)	31	X	X	X	X	X	X	X	X	X

MONTH: J F M A M J J A S O N D

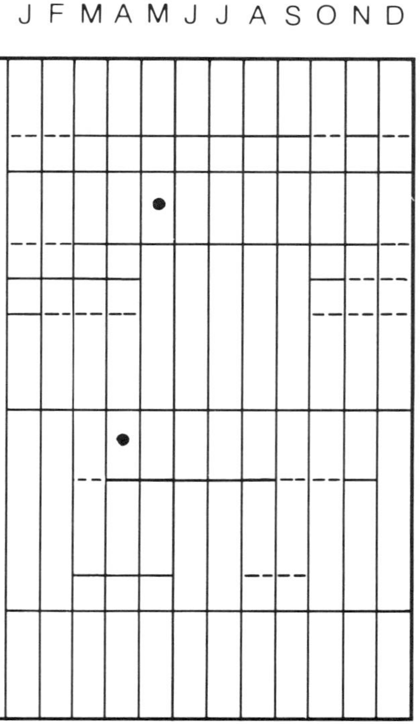

SPECIES		Body Mass (g)	Valley Floor	Human Habitats	Alluvial Plain	Rocky Slopes	Lower Plateau	Piñon-Juniper	Chaparral	Coniferous Forest	Streamside	J	F	M	A	M	J	J	A	S	O	N	D
TROCHILIDAE																							
N	Black-chinned Hummingbird (*Archilochus alexandri*)	3.3		X	X		X				X												
N	Costa's Hummingbird (*Calypte costae*)	3.2	X	X	X	X	X	X	X		X												
N	Anna's Hummingbird (*Calypte anna*)	4.2	X	X	X	X	X	X	X	X	X												
	Rufous Hummingbird (*Selasphorus rufus*)	3.8	X		X	X	X		X	X	X												
	Allen's Hummingbird (*Selasphorus sasin*)	3.7	X		X				X	X	X												
N	Calliope Hummingbird (*Stellula calliope*)	2.6					X		X	X													
ALCEDINIDAE																							
	Belted Kingfisher (*Megaceryle alcyon*)	150	X	X																			
PICIDAE																							
N	Common Flicker (*Colaptes auratus*)	150	X	X	X		X	X	X	X	X												
N	Acorn Woodpecker (*Melanerpes formicivorus*)	80							X	X	X												
	Lewis' Woodpecker (*Melanerpes lewis*)	98						X	X	X													
?N	Yellow-bellied Sapsucker (*Sphyrapicus varius*)	46						X		X	X												
	Williamson's Sapsucker (*Sphyrapicus thyroideus*)	50						X		X											•		•

SPECIES	Body Mass (g)	Valley Floor	Human Habitats	Alluvial Plain	Rocky Slopes	Lower Plateau	Piñon-Juniper	Chaparral	Coniferous Forest	Streamside	MONTH J F M A M J J A S O N D
N Hairy Woodpecker (*Picoides villosus*)	72						X	X	X	X	
N Ladder-backed Woodpecker (*Picoides scalaris*)	36			X	X	X	X	X		X	
N Nuttall's Woodpecker (*Picoides nuttallii*)	35					X	X		X	X	
N White-headed Woodpecker (*Picoides albolarvatus*)	62							X	X	X	
TYRANNIDAE											
N Western Kingbird (*Tyrannus verticalis*)	37	X	X	X	X	X	X	X			
Cassin's Kingbird (*Tyrannus vociferans*)	46				X	X					
Wied's Crested Flycatcher (*Myiarchus tyrannulus*)	–			X							•
N Ash-throated Flycatcher (*Myiarchus cinerascens*)	29	X	X	X	X	X	X	X	X	X	
?N Black Phoebe (*Sayornis nigricans*)	18		X	X						X	
N Say's Phoebe (*Sayornis saya*)	21	X	X	X	X	X		X	X	X	
Willow Flycatcher (*Empidonax traillii*)	12						X				
Hammond's Flycatcher (*Empidonax hammondii*)	10				X	X	X	X	X	X	
N Dusky Flycatcher (*Empidonax oberholseri*)	11							X	X		
?N Gray Flycatcher (*Empidonax wrightii*)	12						X		X		
?N Western Flycatcher (*Empidonax difficilis*)	11			X	X	X	X	X	X	X	
N Western Wood Pewee (*Contopus sordidulus*)	12		X	X	X	X	X	X	X	X	

SPECIES	Body Mass (g)	Valley Floor	Human Habitats	Alluvial Plain	Rocky Slopes	Lower Plateau	Piñon-Juniper	Chaparral	Coniferous Forest	Streamside
N Olive-sided Flycatcher (*Nuttallornis borealis*)	32		X	X				X	X	X
Vermilion Flycatcher (*Pyrocephalus rubinus*)	–			X						
ALAUDIDAE										
Horned Lark (*Eremophila alpestris*)	35	X	X							
HIRUNDINIDAE										
N Violet-green Swallow (*Tachycineta thalassina*)	15	X	X	X	X	X	X	X	X	X
Tree Swallow (*Iridoprocne bicolor*)	18	X	X	X	X	X			X	X
N Rough-winged Swallow (*Stelgidopteryx ruficollis*)	15	X	X	X	X	X		X	X	X
?N Barn Swallow (*Hirundo rustica*)	19	X	X	X	X	X	X	X		X
?N Cliff Swallow (*Petrochelidon pyrrhonota*)	21	X	X	X	X	X				X
CORVIDAE										
N Steller's Jay (*Cyanocitta stelleri*)	110								X	X
N Scrub Jay (*Aphelocoma coerulescens*)	76			X		X	X	X	X	X
N Common Raven (*Corvus corax*)	864	X	X	X	X	X	X	X	X	X
N Piñon Jay (*Gymnorhinus cyanocephalus*)	101						X			
Clark's Nutcracker (*Nucifraga columbiana*)	144								X	

SPECIES	Body Mass (g)	Valley Floor	Human Habitats	Alluvial Plain	Rocky Slopes	Lower Plateau	Piñon-Juniper	Chaparral	Coniferous Forest	Streamside
PARIDAE										
N Mountain Chickadee (*Parus gambeli*)	11.5						X	X	X	X
N Plain Titmouse (*Parus inornatus*)	14					X	X	X	X	X
N Verdin (*Auriparus flaviceps*)	6.5	X	X	X	X	X	X			X
N Bushtit (*Psaltriparus minimus*)	5.0			X		X	X	X	X	X
SITTIDAE										
N White-breasted Nuthatch (*Sitta carolinensis*)	17						X	X	X	X
Red-breasted Nuthatch (*Sitta canadensis*)	10						X		X	
N Pygmy Nuthatch (*Sitta pygmaea*)	10								X	X
CERTHIIDAE										
N Brown Creeper (*Certhia familiaris*)	8.0								X	X
CHAMAEIDAE										
N Wrentit (*Chamaea fasciata*)	16						X	X	X	X
CINCLIDAE										
Dipper (*Cinclus mexicanus*)	50									X

SPECIES	Body Mass (g)	Valley Floor	Human Habitats	Alluvial Plain	Rocky Slopes	Lower Plateau	Piñon-Juniper	Chaparral	Coniferous Forest	Streamside
TROGLODYTIDAE										
N House Wren (*Troglodytes aedon*)	9.8			X			X	X	X	X
N Bewick's Wren (*Thryomanes bewickii*)	9	X		X	X	X	X	X	X	X
N Cactus Wren (*Campylorhynchus brunneicapillus*)	40	X	X	X	X	X	X			X
N Cañon Wren (*Catherpes mexicanus*)	11				X	X	X	X		X
N Rock Wren (*Salpinctes obsoletus*)	16	X	X	X	X	X	X	X	X	X
MIMIDAE										
N Mockingbird (*Mimus polyglottos*)	48	X	X	X		X	X	X		X
N California Thrasher (*Toxostoma redivivum*)	90			X		X	X	X		X
N Le Conte's Thrasher (*Toxostoma lecontei*)	63	X		X						
N Crissal Thrasher (*Toxostoma dorsale*)	60			X						
Sage Thrasher (*Oreoscoptes montanus*)	50	X	X	X						
TURDIDAE										
N American Robin (*Turdus migratorius*)	80		X	X			X	X	X	X
Varied Thrush (*Ixoreus naevius*)	78						X			X
Hermit Thrush (*Catharus guttatus*)	26		X							

MONTH: J F M A M J J A S O N D

SPECIES	Body Mass (g)	Valley Floor	Human Habitats	Alluvial Plain	Rocky Slopes	Lower Plateau	Piñon-Juniper	Chaparral	Coniferous Forest	Streamside	MONTH J F M A M J J A S O N D
?N Swainson's Thrush (*Catharus ustulata*)	27			X			X			X	
N Western Bluebird (*Sialia mexicana*)	27	X	X	X	X	X	X	X	X	X	
Mountain Bluebird (*Sialia currucoides*)	29		X	X			X		X		
N Townsend's Solitaire (*Myadestes townsendi*)	32			X			X		X	X	
SYLVIIDAE											
N Blue-gray Gnatcatcher (*Polioptila caerulea*)	5.5	X		X	X	X	X	X			
N Black-tailed Gnatcatcher (*Polioptila melanura*)	5.2	X	X	X		X				X	
?N Ruby-crowned Kinglet (*Regulus calendula*)	6.8	X	X	X	X	X	X	X	X	X	
MOTACILLIDAE											
Water Pipit (*Anthus spinoletta*)	20	X	X	X							
BOMBYCILLIDAE											
Cedar Waxwing (*Bombycilla cedrorum*)	33		X	X			X				
PTILOGONATIDAE											
N Phainopepla (*Phainopepla nitens*)	24	X	X	X	X	X	X	X		X	

SPECIES	Body Mass (g)	Valley Floor	Human Habitats	Alluvial Plain	Rocky Slopes	Lower Plateau	Piñon-Juniper	Chaparral	Coniferous Forest	Streamside	MONTH J F M A M J J A S O N D
LANIIDAE											
N Loggerhead Shrike (*Lanius ludovicianus*)	46	X	X	X	X	X	X	X		X	
STURNIDAE											
N Starling (*Sturnus vulgaris*)	75	X	X	X			X				
VIREONIDAE											
Bell's Vireo (*Vireo bellii*)	9.9			X			X			X	
N Gray Vireo (*Vireo vicinior*)	12.5				X	X	X	X		X	
N Solitary Vireo (*Vireo solitarius*)	14			X					X	X	
N Warbling Vireo (*Vireo gilvus*)	12	X		X	X	X	X	X	X	X	
PARULIDAE											
N Orange-crowned Warbler (*Vermivora celata*)	9.0	X	X	X	X	X	X	X	X	X	
Nashville Warbler (*Vermivora ruficapilla*)	8.1	X		X	X	X	X	X	X	X	
Yellow Warbler (*Dendroica petechia*)	9.5		X	X			X			X	
N Yellow-rumped Warbler (*Dendroica coronata*)	12	X	X	X	X	X	X	X	X	X	
N Black-throated Gray Warbler (*Dendroica nigrescens*)	8.4			X	X	X	X	X	X	X	

SPECIES	Body Mass (g)	Valley Floor	Human Habitats	Alluvial Plain	Rocky Slopes	Lower Plateau	Piñon-Juniper	Chaparral	Coniferous Forest	Streamside	MONTH J F M A M J J A S O N D
Townsend's Warbler (*Dendroica townsendi*)	8.5	X		X	X	X	X	X	X	X	
Hermit Warbler (*Dendroica occidentalis*)	9.3					X	X	X	X	X	
MacGillivray's Warbler (*Oporornis tolmiei*)	11	X		X			X		X	X	
Common Yellowthroat (*Geothlypis trichas*)	10.5						X			X	
Yellow-breasted Chat (*Icteria virens*)	27			X							
Wilson's Warbler (*Wilsonia pusilla*)	7.2	X	X	X	X	X	X	X	X	X	
PLOCEIDAE											
N House Sparrow (*Passer domesticus*)	26	X	X	X			X			X	
ICTERIDAE											
N Western Meadowlark (*Sturnella neglecta*)	98	X	X	X		X	X				
Yellow-headed Blackbird (*Xanthocephalus xanthocephalus*)	74		X	X							
Red-winged Blackbird (*Agelaius phoeniceus*)	54		X	X			X				
N Hooded Oriole (*Icterus cucullatus*)	25		X	X		X	X			X	
N Scott's Oriole (*Icterus parisorum*)	38	X		X	X	X	X	X		X	
N Northern Oriole (*Icterus galbula*)	34	X	X	X	X	X		X	X	X	
N Brewer's Blackbird (*Euphagus cyanocephalus*)	66	X	X	X				X			

	SPECIES	Body Mass (g)	Valley Floor	Human Habitats	Alluvial Plain	Rocky Slopes	Lower Plateau	Piñon-Juniper	Chaparral	Coniferous Forest	Streamside	MONTH J F M A M J J A S O N D
N	Brown-headed Cowbird (*Molothrus ater*)	35	X	X	X		X		X			
	THRAUPIDAE											
N	Western Tanager (*Piranga ludoviciana*)	30	X	X	X	X	X	X	X	X	X	
	FRINGILLIDAE											
	Rose-breasted Grosbeak (*Pheucticus ludovicianus*)	44		X	X							
N	Black-headed Grosbeak (*Pheucticus melanocephalus*)	45	X	X	X	X	X	X	X	X	X	
	Blue Grosbeak (*Guiraca caerulea*)	–			X							
?N	Lazuli Bunting (*Passerina amoena*)	14			X	X	X	X	X		X	
N	Purple Finch (*Carpodacus purpureus*)	21						X		X	X	
N	Cassin's Finch (*Carpodacus cassinii*)	27						X	X	X	X	
N	House Finch (*Carpodacus mexicanus*)	20	X	X	X	X	X	X	X	X	X	
	Pine Siskin (*Carduelis pinus*)	11				X	X	X	X		X	
	American Goldfinch (*Carduelis tristis*)	13		X	X				X		X	
N	Lesser Goldfinch (*Carduelis psaltria*)	9.0	X	X	X	X	X	X	X	X	X	
N	Lawrence's Goldfinch (*Carduelis lawrencei*)	10						X	X	X	X	

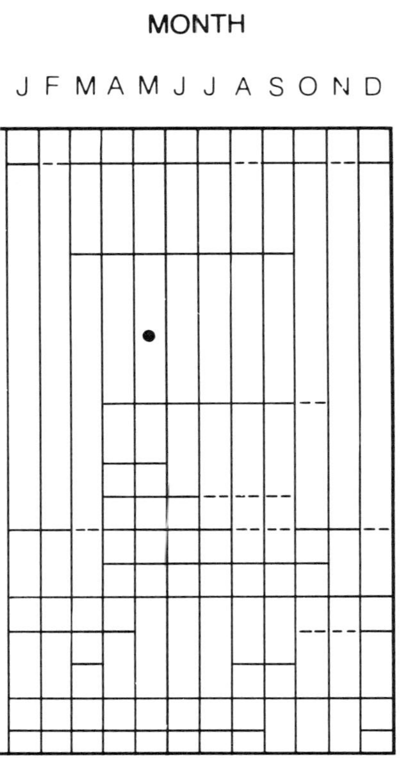

	SPECIES	Body Mass (g)	Valley Floor	Human Habitats	Alluvial Plain	Rocky Slopes	Lower Plateau	Piñon-Juniper	Chaparral	Coniferous Forest	Streamside	J	F	M	A	M	J	J	A	S	O	N	D
	Red Crossbill (*Loxia curvirostra*)	35								X								●					
N	Green-tailed Towhee (*Pipilo chlorura*)	28			X			X	X	X	X												
N	Rufous-sided Towhee (*Pipilo erythrophthalmus*)	38			X			X	X	X	X												
N	Brown Towhee (*Pipilo fuscus*)	44					X	X	X		X												
N	Abert's Towhee (*Pipilo aberti*)	47	X	X	X						X												
	Savannah Sparrow (*Passerculus sandwichensis*)	17	X	X	X					X													
	Vesper Sparrow (*Pooecetes gramineus*)	24	X																				
	Lark Sparrow (*Chondestes grammacus*)	30	X	X	X			X															
	Rufous-crowned Sparrow (*Aimophila ruficeps*)	17	X		X						X			●	●				●			●	
N	Black-throated Sparrow (*Amphispiza bilineata*)	12.6	X	X	X	X	X	X	X		X												
N	Sage Sparrow (*Amphispiza belli*)	15	X		X				X		X												
N	Dark-eyed Junco (*Junco hyemalis*)	17			X		X	X	X	X	X												
N	Chipping Sparrow (*Spizella passerina*)	12	X		X	X	X	X	X	X	X												
?N	Brewer's Sparrow (*Spizella breweri*)	11	X		X		X	X	X	X	X												
N	Black-chinned Sparrow (*Spizella atrogularis*)	11			X				X	X													
	White-crowned Sparrow (*Zonotrichia leucophrys*)	29	X	X	X	X	X	X	X	X	X												
	Golden-crowned Sparrow (*Zonotrichia atricapilla*)	35			X		X	X	X		X												
?N	Fox Sparrow (*Passerella iliaca*)	31						X	X	X	X												
	Lincoln's Sparrow (*Melospiza lincolnii*)	17									X									●			

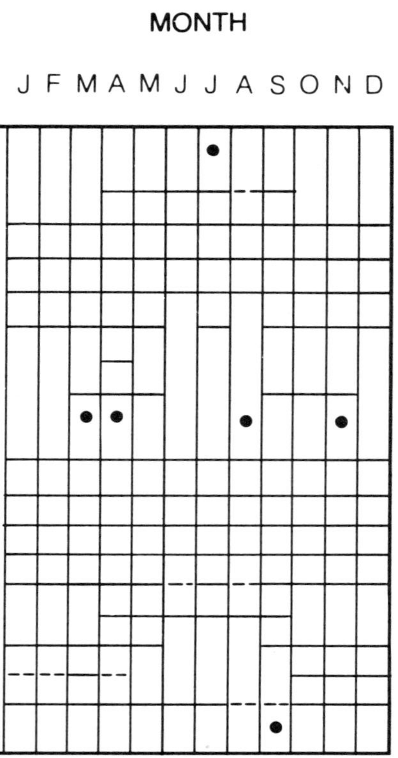

REFERENCES

AMERICAN ORNITHOLOGISTS' UNION. 1957. *Check-list of North American birds*, 5th ed. Baltimore, Md.: American Ornithologists' Union.

ANDERSON, A. H., and A. ANDERSON. 1963. Life history of the Cactus Wren. Part IV. Competition and survival. *Condor* 65: 29–43.

———. 1973. *The Cactus Wren*. Tucson: University of Arizona Press.

ANDERSON, B. W., and R. D. OHMART. 1977. *Climatological and physical characteristics affecting avian population estimates in southwestern riparian communities using transect counts*. U. S. For. Serv. Gen. Tech. Rep. RM-43, pp. 193–200.

AUSTIN, G. T. 1978. Daily time budget of the postbreeding Verdin. *Auk* 95: 247–251.

AUSTIN, G. T., and E. L. SMITH. 1974. Use of burrows by Brown Towhees and Black-throated Sparrows. *Auk* 91: 167.

AXELROD, D. I. 1973. History of the Mediterranean ecosystem in California. Pp. 225–227. In F. di Castri and H. A. Mooney, eds. *Mediterranean type ecosystems, origin and structure*. New York: Springer-Verlag.

BAILEY, A. M., and R. J. NIEDRACH. 1965. *Birds of Colorado*. Vol. 2. Denver: Denver Museum of Natural History.

BARBOUR, M. G., and J. MAJOR. 1977. *Terrestrial vegetation of California*. New York: John Wiley and Sons.

BARTHOLOMEW, G. A. 1972. The water economy of seed-eating birds that survive without drinking. Pp. 237–254. In K. H. Voous, ed. Brill, Leiden: *Proc. XVth Internat. Ornith. Congr.*

BARTHOLOMEW, G. A., and T. J. CADE. 1956. Water consumption of House Finches. *Condor* 58: 406–412.

———. 1963. The water economy of land birds. *Auk* 80: 504–539.

BARTHOLOMEW, G. A., T. R. HOWELL, and T. J. CADE. 1957. Torpidity in the White-throated Swift, Anna Hummingbird, and Poor-will. *Condor* 59: 145–155.

BEAL, F. E. L. 1911. *Food of the woodpeckers of the United States*. Washington, D.C.; U.S. Dept. Agric., Bur. Biol. Bull. no. 37.

BEHLE, W. H. 1978. Avian biogeography of the Great Basin and Intermountain Region. Pp. 55–76. In K. T. Harper and J. L. Reveal, eds. *Intermountain biogeography: a symposium*. Great Basin Naturalist Memoirs, no. 2. Provo, Utah: Brigham Young University.

BELDING, L. 1890. Land birds of the Pacific district. *Occ. Papers Calif. Acad. Sci.* 2: 1–274.

BENT, A. C. 1932. *Life histories of North American gallinaceous birds.* U. S. Natl. Mus., Bull. 162. Washington, D. C.: Smithsonian Institution.

———. 1937a. *Life histories of North American birds of prey, part 1.* U. S. Natl. Mus., Bull. 167. Washington, D. C.: Smithsonian Institution.

———. 1937b. *Life histories of North American birds of prey, part 2.* U. S. Natl. Mus., Bull. 170. Washington, D. C.: Smithsonian Institution.

———. 1940. *Life histories of North American cuckoos, goatsuckers, hummingbirds and their allies, part 1.* U. S. Natl. Mus., Bull 176. Washington, D. C.: Smithsonian Institution.

———. 1948. *Life histories of North American nuthatches, wrens, thrashers, and their allies.* U. S. Natl. Mus., Bull. 195. Washington, D. C.: Smithsonian Institution.

BORROR, D. J. 1977. Rufous-sided Towhees mimicking Carolina Wren and Field Sparrow. *Wilson Bull.* 89: 477–480.

BROWN, J. H., and D. W. DAVIDSON. 1977. Competition between seed-eating rodents and ants in desert ecosystems. *Science* 196: 880–882.

CALDER, W. A. 1968. Nest sanitation: A possible factor in the water economy of the roadrunner. *Condor* 70: 279.

———. 1974. Consequences of body size for avian energetics. Pp. 86–144. In R. A. Paynter, Jr., ed. *Avian energetics.* Cambridge, Mass.: Publ. Nuttall Ornithol. Club, no. 15.

CALDER, W. A., and K. SCHMIDT-NIELSEN. 1967. Temperature regulation and evaporation in the pigeon and roadrunner. *Amer. J. Physiol.* 213: 883–889.

CAREY, C., and M. L. Morton. 1971. A comparison of salt and water regulation in California Quail (*Lophortyx californicus*) and Gambel's Quail (*Lophortyx gambelli*). *Comp. Biochem. Physiol.* 39A: 75–101.

CARLSON, B. A. 1979a. Ocotillo—cholla. *Amer. Birds* 33: 36–37.

———. 1979b. Ocotillo—cholla. *Amer. Birds* 33: 94.

CARPENTER, F. L., and J. L. CASTRONOVA. 1980. Material diet selectivity in *Calypte anna. Amer. Midland Nat.* 103: 175–179.

Cody, M. L. 1971. Finch flocks in the Mojave desert. *Theor. Popul. Biol.* 2: 142–158.

———. 1973. Parallel evolution in bird niches. Pp. 307–338. In F. di Castri and H. A. Mooney, eds. *Mediterranean type ecosystems, origin and structure.* New York: Springer-Verlag.

———. 1974. *Competition and the structure of bird communities.* Princeton: Princeton University Press.

COGSWELL, H. L. 1977. *The water birds of California.* Berkeley, Los Angeles, and London: University of California Press.

CORD, B., and J. R. JEHL, JR. 1979. Distribution, biology, and status of a relict population of Brown Towhee (*Pipilo fuscus eremophilus*). *West. Birds* 10: 131–156.

COULOMBE, H. N. 1970. Physiological and physical aspects of temperature regulation in the Burrowing Owl, *Speotyto cunicularia. Comp. Biochem. Physiol.* 35: 307–337.

———. 1971. Behavior and population ecology of the Burrowing Owl, *Speotyto cunicularia*, in the Imperial Valley of California. *Condor* 73: 162–176.

DAWSON, W. L. 1923. *The birds of California.* San Diego, Calif: South Moulton Co.

DAWSON, W. R. 1954. Temperature regulation and water requirements of the Brown and Abert Towhees, *Pipilo fuscus* and *Pipilo aberti. Univ. Calif. (Berkeley) Publ. Zool.* 59: 81–124.

———. 1976. Physiological and behavioral adjustments of birds to heat and aridity. Pp. 455–467. In H. J. Firth and J. H. Calaby, eds. *Proc. XVIth Internat. Ornith. Congr.* Canberra City: Australian Academy of Science.

Dawson, W. R., and G. A. Bartholomew. 1968. Temperature regulation and water economy of desert birds. Pp. 357–394. In G. W. Brown, ed. *Desert biology*, vol. I. New York: Academic Press.

Dawson, W. R., C. Carey, C. S. Adkisson, and R. D. Ohmart. 1979. Responses of Brewer's and Chipping Sparrows to water restriction. *Physiol. Zool.* 52: 529–541.

Dickson, J. G., R. N. Connor, R. R. Fleet, J. A. Jackson, and J. C. Kroll. 1979. *The role of insectivorous birds in forest ecosystems.* New York: Academic Press.

Dunn, J. 1977. The genus *Empidonax. West. Tanager* 43: 5–7.

Dunn, J., and P. Unitt. 1977. A Laysan Albatross in interior southern California. *West. Birds* 8: 27–28.

Emlen, J. T. 1979. Land bird densities on Baja California islands. *Auk* 96: 152–167.

Engles, W. 1940. Structural adaptations in thrashers (Mimidae: genus *Toxostoma*) with comments on interspecific relationships. *Univ. Calif. (Berkeley) Publ. Zool.* 42: 341–400.

Erickson, M. M. 1948. Gambel's Wren-tit. Pp. 81–93. In A. C. Bent, ed. *Life histories of North American nuthatches, wrens, thrashers, and their allies.* U. S. Natl. Mus., Bull. 195. Washington, D. C.: Smithsonian Institution.

Fox, R., S. W. Lehmkuhle, and D. H. Westendorf. 1976. Falcon visual acuity. *Science* 192: 263–265.

Franzreb, K. E. 1978. Breeding bird densities, species composition, and bird species diversity of the Algodones dunes. *West. Birds* 9: 9–20.

Friedmann, H. 1950. *The birds of North and Middle America.* Washington, D. C.: U. S. Natl. Mus., Bull. no. 50.

Furness, R. W. 1978. Energy requirements of seabird communities: a bioenergetics model. *J. Animal Ecol.* 47: 39–53.

Garrett, K., and J. Dunn. 1981. *Birds of Southern California status and distribution.* Los Angeles: Los Angeles Audubon Soc.

Gates, D. M., and R. B. Schmerl. 1975. *Perspectives of biophysical ecology.* New York: Springer-Verlag.

Goldsmith, T. H., and K. M. Goldsmith. 1979. Discrimination of colors by the Black-chinned Hummingbird, *Archilochus alexandri. J. Comp. Physiol.* 130: 209–220.

Goodwin, G. A. 1977. Golden Eagle predation on pronghorn antelope. *Auk* 94: 789–790.

Gordon, S. H. 1978. "Food and foraging ecology of a desert harvester ant, Veromessor pergandei (Mayr)." Ph.D. diss. University of California, Berkeley.

Grant, K. A., and V. Grant. 1968. *Hummingbirds and their flowers.* New York: Columbia University Press.

Grinnell, J. 1904. Midwinter birds at Palm Springs, California. *Condor* 6: 40–45.

Grinnell, J., J. Dixon, and J. M. Linsdale. 1930. Vertebrate natural history of a section of northern California through the Lassen Park region. *Univ. Calif. (Berkeley) Publ. Zool.* 35: 1–594.

Grinnell, J., and A. H. Miller. 1944. The distribution of birds of California. *Pac. Coast Avifauna*, no. 27: 1–608.

GRINNELL, J., and H. S. SWARTH. 1913. An account of the birds and mammals of the San Jacinto area of southern California with remarks upon the behavior and geographic races on the margins of their habitats. *Univ. Calif. (Berkeley) Publ. Zool.*, 10: 197–406.

GRUSON, E. S. 1972. *Words for birds*. New York: Quadrangle Books.

HAILS, C. J. 1979. A comparison of flight energetics in hirundines and other birds. *Comp. Biochem. Physiol.* 63A: 581–585.

HARRISON, C. 1978. *A field guide to the nests, eggs and nestlings of North American birds*. Glasgow: William Collin Sons & Co., Ltd.

HENNY, C. J. 1972. *An analysis of population dynamics of selected avian species with special reference to the modern pesticide era*. U. S. Fish Wildl. Serv., Wildl. Res. Rep. 1.

HOFFMANN, R. 1955. *Birds of the Pacific states*. Boston: Houghton Mifflin Co.

HOLMES, R. T., R. E. BONNEY, JR., and S. W. PACALA. 1979. Guild structure of the Hubbard Brook bird community: a multivariate approach. *Ecology* 60: 512–520.

HOLMES, R. T., and F. W. STURGES. 1975. Avian community dynamics and energetics in a northern hardwoods ecosystem. *J. Animal Ecol.* 44: 175–200.

HUBBARD, J. P. 1974. Avian evolution in the aridlands of North America. *Living Bird* 12: 155–196.

JAEGER, E. C. 1948. Does the Poor-will hibernate? *Condor* 50: 45–46.

———. 1957. *The North American deserts*. Stanford: Stanford University Press.

———. 1965. *The California deserts*, 4th ed. Stanford: Stanford University Press.

JOHNSON, N. K., and K. L. GARRETT. 1974. Interior bird species expand breeding ranges into southern California. *West. Birds* 5: 45–56.

KENDEIGH, S. C., and J. PINOWSKI. 1973. *Productivity, population dynamics and systematics of granivorous birds*. Warsaw: PWN—Polish Scientific Publishers.

KOSKIMIES, J. 1948. On temperature regulation and metabolism in the swift, *Micropus a. apus* L., during fasting. *Experimentia* 4: 274–276.

KNIGHT, F. B. 1958. The effects of woodpeckers on populations of the Engelmann spruce beetle. *J. Econ. Entomol.* 51: 603–607.

KNORR, O. A. 1957. Communal roosting of the Pygmy Nuthatch. *Condor* 59: 398.

KREBS, C. J. 1978. *Ecology: The experimental analysis of distribution and abundance*. New York: Harper and Row.

LACK, D. L. 1954. *The natural regulation of animal numbers*. Oxford: Clarendon Press.

LASIEWSKI, R. C. 1964. Body temperature, heart and breathing rate, and evaporative water loss in hummingbirds. *Physiol. Zool.* 37: 212–223.

LAWRENCE, L. de K. 1967. *A comparative life-history study of four species of woodpeckers*. Baltimore, Md.: Monogr. no. 5. Amer. Ornith. Union.

LEOPOLD, A. S., M. ERWIN, J. Oh, and B. BROWNING. 1976. Phytoestrogens: Adverse effects on reproduction in California Quail. *Science* 191: 98–100.

LIGON, J. D. 1973. Foraging behavior of the White-headed Woodpecker in Idaho. *Auk* 90: 862–869.

———. 1978. Reproductive interdependence of Piñon Jays and Piñon Pines. *Ecol. Monogr.* 48: 111–126.

LOWE, C. H. 1964. *Arizona's natural environment*. Tucson: University of Arizona Press.

MACLEAN, G. L. 1974. Arid-zone adaptations in southern African birds. *Cimbebasia (A)* 3: 163–175.

MACROBERTS, B. R., and M. H. MACROBERTS. 1976. *Social organization and behavior of the Acorn Woodpecker in central coastal California.* Baltimore, Md.: Monogr. no. 21. Amer. Ornith. Union.

MALLETTE, R. D., and G. I. GOULD, JR. 1976. *Raptors of California.* Sacramento, Ca.: California Dept. Fish and Game.

MARSHALL, J. T., JR. 1960. Interrelations of Abert and Brown Towhees. *Condor* 62: 49–64.

MARTIN, A. C., H. S. ZIM, and A. L. NELSON. 1951. *American wildlife & plants: A guide to wildlife food habits.* New York: McGraw-Hill Book Co.

MAYR, E. 1964*a*. Nearctic region. Pp. 514–516. In A. L. Thompson, ed. *A new dictionary of birds.* New York: McGraw Hill.

———. 1964*b*. Inferences concerning the Tertiary American bird faunas. *Proc. Natl. Acad. Sci.* 51: 280–288.

MAYR, E., and L. L. SHORT. 1970. *Species taxa of North American birds.* Cambridge, Mass.: Publ. Nuttall Ornith. Club, no. 9.

MCGINNIES, W. G., B. J. GOLDMAN, and P. PAYLORE. 1968. *Deserts of the world.* Tucson: University of Arizona Press.

MEAD, C. J. 1978. Old World Warblers. Pp. 213–214. In C. J. O. Harrison, ed. *Bird families of the world.* New York: Abrams.

MEBS, T. 1972. Family: Falcons. Pp. 411–431. In B. Grizmek, ed. *Grizmek's animal life encyclopedia,* vol. 7. New York: Van Nostrand-Reinhold.

MEIGS, P. 1953. World distribution of arid and semi-arid homoclimates. Pp. 203–210. In *Reviews of research on arid zone hydrology.* Paris: UNESCO.

MERRIAM, C. H. 1898. *Life-zones and crop-zones of the United States.* U. S. Dept. of Agric., Div. Biol. Surv., Bull. 10: 1–79.

MILLER, A. H. 1931. Systematic revision and natural history of the American shrikes (*Lanius*). *Univ. Calif. (Berkeley) Publ. Zool.* 38: 11–242.

———. 1936. Tribulations of thorn-dwellers. *Condor* 38: 218–219.

———. 1951. An analysis of the distribution of the birds of California. *Univ. Calif. (Berkeley) Publ. Zool.* 50: 531–644.

MILLER, A. H., and R. C. STEBBINS. 1964. *The lives of desert animals in Joshua Tree National Monument.* Berkeley and Los Angeles: University of California Press.

MOLDENHAUER, R. R., and P. G. TAYLOR. 1973. Energy intake by hydropenic Chipping Sparrows (*Spizella passerina passerina*) maintained on different diets. *Condor* 75: 439–445.

MOLDENHAUER, R. R., and J. A. WIENS. 1970. The water economy of the Sage Sparrow, *Amphispiza belli nevadensis. Condor* 72: 265–275.

MOONEY, H. A., and E. L. DUNN. 1970. Convergent evolution of Mediterranean-climate evergreen sclerophyll shrubs. *Evol.* 24: 292–303.

MOONEY, H. A., and D. J. PARSONS. 1973. Structure and function of the California chaparral—an example from San Dimas. Pp. 83–112. In F. di Castri and H. A. Mooney, eds. *Mediterranean type ecosystems, origin and structure.* New York: Springer-Verlag.

MORSE, D. H. 1968. The use of tools by Brown-headed Nuthatches. *Wilson Bull.* 80: 220–224.

OHMART, R. D. 1972. Salt-secreting nasal gland and its ecological significance in the roadrunner. *Comp. Biochem. Physiol.* 43A: 311–316.

———. 1973. Observations on the breeding adaptations of the roadrunner. *Condor* 75: 140–149.

OHMART, R. D., and R. C. LASIEWSKI. 1971. Roadrunners: Energy conservation by hypothermia and absorption of sunlight. *Science* 172: 67–69.

PAYNE, R. B. 1965. Clutch size and number of eggs laid by Brown-headed Cowbird. *Condor* 67: 44–60.

PETERSON, R. T. 1961. *A field guide to western birds*, 2d ed. Boston: Houghton Mifflin Co.

PHILLIPS, A., J. MARSHALL, and G. MONSON. 1964. *The birds of Arizona*. Tucson: University of Arizona Press.

PIANKA, E. R. 1978. *Evolutionary ecology*, 2d ed. New York: Harper and Row.

PIELOU, E. C. 1974. *Population and community ecology principles and methods*. New York: Gordon and Breach Science Publishers.

PINOWSKI, J., and S. C. KENDEIGH. 1977. *Granivorous birds in ecosystems*. Cambridge: Cambridge Univ. Press.

PLUMB, T. R. 1961. *Sprouting of chaparral by December after a wildfire in July*. U. S. For. Ser. Pacific Southwest Forest and Range Expt. Sta. Tech. Paper 57.

RAPHAEL, M. G. 1980. "Utilization of standing dead trees by breeding birds at Sagehen Creek, California." Ph.D. diss. University of California, Berkeley.

RENOLDS, T. D. 1979. The impact of Loggerhead Shrikes on nesting birds in a sagebrush environment. *Auk* 96: 798–799.

RICKLEFS, R. E. 1973. *Ecology*. Portland, Oregon: Chiron Press.

RICKLEFS, R. E., and F. R. HAINSWORTH. 1968. Temperature dependent behavior of the Cactus Wren. *Ecol.* 49: 227–233.

———. 1969. Temperature regulation in the nestling Cactus Wren: The nest environment. *Condor* 71: 32–37.

ROTENBERRY, J. T., R. E. FITZER, and W. H. RICKARD. 1979. Seasonal variation in avian community structure: differences in mechanisms regulating diversity. *Auk* 96: 499–505.

RYAN, R. M. 1968. *Mammals of Deep Canyon Colorado Desert, California*. Palm Springs, Ca.: Desert Museum.

SARGENT, G. I. 1940. Observations on color-banded California Thrashers. *Condor* 42: 49–60.

SHANTZ, H. L. and R. L. PIEMEISEL. 1924. Indicator significance of the natural vegetation of the southwestern desert region. *J. Agric. Res.* 28: 721–801.

SHORT, L. L. 1969. Taxonomic aspects of avian hybridization. *Auk* 86: 84–105.

———. 1971. Systematics and behavior of some North American woodpeckers, genus *Picoides* (Aves). *Bull. Amer. Mus. Nat. Hist.* 145: 5–118.

SHREVE, F. 1925. Ecological aspects of the deserts of California. *Ecology* 6: 93–103.

SILLMAN, A. J. 1973. Avian Vision. Pp. 349–387. In D. S. Farner and J. R. King, eds. *Avian Biology*, vol. 3. New York: Academic Press.

SMALL, A. 1974. *The birds of California*. New York: Winchester Press.

SMYTH, M., and G. A. BARTHOLOMEW. 1966. The water economy of the Black-throated Sparrow and the Rock Wren. *Condor* 68: 447–458.

STEINBACHER, K. 1964. Woodpeckers. Pp. 895–897. In A. L. Thompson, ed. *A new dictionary of birds*. New York: McGraw Hill.

STILES, F. G. 1973. Food supply and the annual cycle of the Anna Hummingbird. *Univ. Calif. (Berkeley) Publ. Zool.* 97: 1–109.

———. 1976. Taste preferences, color preferences and flower choice in hummingbirds. *Condor* 78:10–26.

Szaro, R. C., and R. P. Balda. 1979. *Bird community dynamics in a ponderosa pine forest.* Studies in Avian Biology, no. 3. Los Angeles: Cooper Ornithol. Soc.

Taylor, W. K. 1971. A breeding biology study of the Verdin, *Auriparus flaviceps* (Sundevall) in Arizona. *Amer. Midland Nat.* 85: 289–328.

Tevis, L., Jr. 1953. Effect of vertebrate animals on seed crop of sugar pine. *J. Wildl. Manage.* 17: 128–131.

Tordoff, H. B., and W. R. Dawson. 1965. The influence of daylength on reproductive timing in the Red Crossbill. *Condor* 67: 416–422.

Tucker, V. A., and G. C. Parrott. 1970. Aerodynamics of gliding flight in a falcon and other birds. *J. Exptl. Biol.* 52: 345–367.

Van Devender, T., and W. G. Spaulding. 1979. Development of vegetation and climate in the southwestern United States. *Science* 204: 701–710.

Van Tyne, J., and A. J. Berger. 1976. *Fundamentals of ornithology*. New York: John Wiley & Sons, Inc.

Walsberg, G. E. 1975. Digestive adaptations of *Phainopepla nitens* associated with the eating of mistletoe berries. *Condor* 77: 169–174.

———. 1977. Ecology and energetics of contrasting social systems in *Phainopepla nitens* (Aves: Ptilogonatidae). *Univ. Calif. Publ. Zool.* 108: 1–63.

———. 1980. Energy expenditure in free living birds: patterns and diversity. Pp. 300–305. In R. Nohring, ed. *Acta XVII Congress Internationalis Ornothologici.* Berlin: Deutsche Ornithologen-Gesellschaft.

Walsberg, G. E., G. S. Campbell, and J. R. King. 1978. Animal coat color and radiative heat gain: A re-evaluation. *J. Comp. Physiol.* 126: 211–222.

Walter, H. 1971. *Ecology of tropical and subtropical vegetation.* New York: Van Nostrand-Reinhold.

Weathers, W. W. 1981. Physiological thermoregulation in heat stressed birds: Consequences of body size. *Physiol. Zool.* 54: 345–361.

Weathers, W. W., and W. W. Mayhew. 1981. Time of day and desert bird censuses. *Western Birds.* 12:157–171.

Weathers, W. W., and K. A. Nagy. 1980. Simultaneous doubly labeled water ($^3HH^{18}0$) and time-budget estimates of daily energy expenditure in *Phainopepla nitens. Auk* 97: 861–867.

Weiner, J., and Z. Glowacinski. 1975. Energy flow through a bird community in a deciduous forest in southern Poland. *Condor* 77: 233–242.

Wells, S., R. A. Bradley, and L. F. Baptista. 1978. Hybridization in *Calypte* hummingbirds. *Auk* 95: 537–549.

Wetmore, A. 1964. *Song and garden birds of North America.* Washington, D. C.: National Geographic Society.

Weymouth, R. D., R. C. Lasiewski, and A. J. Berger. 1964. The tongue apparatus in hummingbirds. *Acta Anat.* 58: 252–270.

Wheeler, G. C., and J. Wheeler. 1973. *Ants of Deep Canyon.* Palm Desert, Ca.: Philip L. Boyd Deep Canyon Desert Research Center.

Wiens, J. A., and G. S. Innes. 1974. Estimation of energy flow in bird communities: A population bioenergetics model. *Ecology* 55: 730–746.

Wiens, J. A., and J. M. Scott. 1975. Model estimation of energy flow in Oregon coastal seabird populations. *Condor* 77: 439–452.

WILBUR, S. R. 1979. The Bell's Vireo in California: A preliminary report. *Amer. Birds* 33: 252.

WILKE, P. J. 1976. Ethnography. Pp. 97–105. In I. P. Ting and B. Jennings, eds. *Deep Canyon, a desert wilderness for science*. Riverside: University of California.

WILLIS, E. C. 1963. Is the Zone-tailed Hawk a mimic of the Turkey Vulture? *Condor* 65: 313–317.

WOODS, R. S. 1927. The hummingbirds of California. *Auk* 44: 297–318.

ZABRISKIE, J. G. 1979. *Plants of Deep Canyon and the central Coachella Valley, California*. Palm Desert, Ca.: Philip L. Boyd Deep Canyon Desert Research Center.

ZIMMERMAN, D. A. 1970. Roadrunner predation on passerine birds. *Condor* 72: 475–476.

INDEX

Page numbers referring to species accounts are italicized.

Designer: Linda Robertson
Compositor: Trend Western
Printer: Malloy Lithographing
Binder: Malloy Lithographing
Text: Baskerville
Display: Bodoni
Color Plates: Lehigh Press